ALADDIN'S LAMP

ALADDIN'S LAMP

How Greek Science Came to Europe
Through the Islamic World

JOHN FREELY

Alfred A. Knopf   New York   2009

THIS IS A BORZOI BOOK
PUBLISHED BY ALFRED A. KNOPF

Grateful acknowledgment is made to the following for permission to
reprint previously published material:

Cambridge University Press: Excerpt from a poem by Xenophanes, translated
by William K. C. Guthrie, from *A History of Greek Philosophy, Volume 1* by
William K. C. Guthrie, copyright © 1962 by Cambridge University Press.
Reprinted by permission of Cambridge University Press.

Dutton: Excerpt from *The Romance of the Rose* by Guillaume de Lorris and
Jean de Meun, edited by Charles W. Dunn, translated by Harry W. Robbins,
copyright © 1962 by Florence L. Robbins. Reprinted by permission
of Dutton, a division of Penguin Group (USA) Inc.

H. D. Miller: Excerpt from "The Dove's Neck-Ring" by Ibn Hazm, translated by
H. D. Miller, originally published in *Ornament of the World: How Muslims, Jews and
Christians Created a Culture of Tolerance in Medieval Spain* by Maria Rosa Menocal
(Boston: Little, Brown, 2002). Reprinted by permission of H. D. Miller.

Michael Fishbein and Weidenfeld and Nicolson: Excerpt from a poem by Khuraymi
abridged by Hugh Kennedy and translated by Michael Fishbein from *The Court
of Caliphs: The Rise and Fall of Islam's Greatest Dynasty* by Hugh Kennedy, copyright
© 2004 by Hugh Kennedy. Reprinted by permission of Michael Fishbein and
Weidenfeld and Nicolson, an imprint of the Orion Publishing Group, London.

Owing to limitations of space, all acknowledgments to reprint previously
published illustrations can be found on page 259.

Library of Congress Cataloguing-in-Publication Data
Freely, John.
Aladdin's lamp : how Greek science came to Europe through
the Islamic world / by John Freely.
p. cm.
Includes bibliographical references and index.
ISBN 978-0-307-26534-0
1. Science—Greece—History.   2. Science—Islamic countries—History.
3. Science, Ancient.   4. Greece—History.   5. Islam and science.   I. Title.
Q127.G7.F74 2009
509—dc22   2008042743

Manufactured in the United States of America
Published February 18, 2009
Second Printing, May 2009

To Toots, as always

CONTENTS

LIST OF ILLUSTRATIONS

ALADDIN'S LAMP

INTRODUCTION

Modern science traces its origins back to ancient Greece, beginning with the first philosophers of nature in the sixth century B.C. Greek science flourished for more than a millennium, ending with the collapse of classical civilization in the early Christian era, when virtually all the cities of the Greco-Roman world were utterly destroyed, beginning the Dark Ages of western Europe. And yet a thousand years later Greek classics inspired the Renaissance and brought about the rebirth of science. When Copernicus published his sun-centered planetary theory in 1543, he was reviving the work of a Greek astronomer who had proposed the same idea some eighteen centuries earlier.

How did ancient Greek science survive and by what means was it transmitted to western Europe? The answer to those questions is the main theme of this book, a story that begins on the Aegean shore of Asia Minor at Miletus, where the first Greek physicists emerged, influenced by ancient Mesopotamian lore in astronomy and mathematics. The story then moves in turn to classical Athens, Hellenistic Alexandria, imperial Rome, Byzantine Constantinople, Nestorian Jundishapur, Abbasid Baghdad, Fatimid Cairo and Damascus, Muslim Cordoba, Toledo of the Reconquista, Norman Palermo, and Latin Oxford and Paris in the thirteenth century, setting the stage for the European scientific revolution of the sixteenth and seventeenth centuries, and finally to Mongol Samarkand and Ottoman Istanbul, tracking the last peak of Islamic science and its long decline.

The story has never been fully told in a book for the general reader, and there is not even a specialist work on the subject as a whole, as I learned when I began my career as a physicist and started reading about the history of science. My first studies in this field began in 1966–67, when I had a postdoctoral fellowship at Oxford under the tutelage of Alistair Crombie, who pioneered the study of how Greek science came to western Europe in translation from Arabic into Latin after having been preserved and developed in the Islamic world. This led me to study the Islamic renaissance of the eighth and ninth centuries, when the translation of Greek scientific and philosophical works into Arabic under the Abbasid caliphs in Baghdad began the first stage in the journey that would eventually lead to the emergence of science in Europe. According to Dimitri Gutas of Yale, one of the leading authorities on the transmission of Greek culture to the Islamic world, "the Graeco-Arabic translation movement of Baghdad constitutes a truly epoch-making stage . . . equal in significance . . . to that of Pericles' Athens, the Italian Renaissance, or the scientific revolution of the sixteenth and seventeenth centuries, and it deserves so to be recognized and embedded in our historical consciousness."

This is not an academic work, but a book designed for the general reader with an interest in cultural history off the beaten track. The emphasis throughout is on the people, places, and cultures involved in the story, an intellectual travelogue that makes its way back and forth between East and West with the tides of history and the rise and fall of civilizations.

The multifaceted cultural interaction that has produced modern science should be of particular interest now, in light of the apocalyptic talk of a clash of civilizations between Islam and the West. The original conflict that accompanied the rise of Islam brought Greco-Islamic science to the West, beginning the modern scientific tradition. The time seems to be right for this story to be told, in all of its cultural complexity. As Edward Said remarked of the interconnected world in which this story is set: "Partly because of empire, all cultures are involved in one another, none is single and pure, all are hybrid, heterogeneous, extraordinarily differentiated and unmonolithic."

So here is the story of how Greek science came to Europe through the Islamic world, beginning with the ancient Ionian city of Miletus in the archaic period of Greek history (750–480 B.C.).

IONIA: THE FIRST PHYSICISTS

The site of ancient Miletus is on the Aegean coast of Turkey south of Izmir, the Greek Smyrna. When I first visited Miletus, in April 1961, it was completely deserted except for a goatherd and his flock, whose resonant bells broke the silence enveloping the ruins through which I wandered, the great Hellenistic theater, the cavernous Roman baths, the colonnaded way that led down to the Lion Port and its surrounding shops and warehouses, once filled with goods from Milesian colonies as far afield as Egypt and the Pontus. Its buildings were now utterly devastated and partly covered with earth, from which the first flowers of spring were emerging, blood-red poppies contrasting with the pale white marble remnants of the dead city.

The site has been under excavation since the late nineteenth century, so that all of its surviving monuments have been unearthed and to some extent restored, though its ancient harbor, the Lion Port, has long been silted up, leaving Miletus marooned miles from the sea. The entrance to the port is still guarded by the marble statues of the two couchant lions from which it took its name, though they are now half-buried in alluvial earth, symbols of the illustrious city that Herodotus called "the glory of Ionia." The Greek geographer Strabo writes that "many are the achievements of this city, but the greatest are the number of its colonizations, for the Euxine Pontus [Black Sea] has been colonized everywhere by these people, as has the Propontis [Sea of Marmara] and several other regions."

Excavations have revealed that the earliest remains in Miletus date from the second half of the sixteenth century B.C., when colonists from Minoan Crete are believed to have established a settlement here. A second colony was founded on the same site during the mass migration of Greeks early in the first millennium B.C., when they left their homeland in mainland Greece and migrated eastward across the Aegean, settling on the coast of Asia Minor and its offshore islands. Three Greek tribes were involved in this migration—the Aeolians to the north, the Ionians in the center, and the Dorians in the south—and together they produced the first flowering of Greek culture. The Aeolians gave birth to the lyric poet Sappho; the Ionians to Homer and the natural philosophers Thales, Anaximander, and Anaximenes; and the Dorians to Herodotus, the "Father of History."

Herodotus, describing this migration in Book I of his *Histories,* writes that the Ionians ended up with the best location in Asia Minor, for they "had the good fortune to establish their settlements in a region which enjoys a better climate than any we know of." Pausanias, in his *Description of Greece,* written in the second century A.D. remarks, "The Ionian countryside has excellently tempered seasons, and its sanctuaries are unrivalled." He goes on to say that "the wonders of Ionia are numerous, and not much short of the wonders of Greece."

The Ionian colonies soon organized themselves into a confederation called the Panionic League. This comprised one city each on the islands of Chios and Samos and ten on the mainland of Asia Minor opposite, namely, Phocaea, Clazomenae, Erythrae, Teos, Lebedus, Colophon, Ephesus, Priene, Myus, and Miletus. The confederation, also known as the Dodecapolis, had its common meeting place at the Panionium, on the mainland opposite Samos. The Ionians also met annually on the island of Delos, the legendary birthplace of Apollo, their patron deity. There they honored the god in a festival described in the Homeric Hymn addressed to Delian Apollo:

> Yet in Delos do you most delight your heart; for the long-robed Ionians gather in your honor with their children and shy wives. Mindful, they delight you with boxing and dancing and song, so often as they hold their gathering. A man would say that they are deathless and unaging if he should

come upon the Ionians so met together. For he would see the graces of them all, and would be pleased in heart gazing at the well-girded women and the men with their swift ships and great renown.

Miletus greatly surpassed all of the other Ionian cities in its maritime ventures and commerce, founding its first colonies in the eighth century B.C. on the shores of the Black Sea. During the next two centuries Miletus was far more active in colonization than any other city-state in the Greek world, founding a total of thirty cities around the Black Sea and its approaches in the Hellespont and the Sea of Marmara. Miletus also had a trading station at Naucratis, the Greek emporium on the Nile delta founded circa 650 B.C. Meanwhile other Greek cities had established colonies around the western shores of the Mediterranean, the densest region of settlement being in southern Italy and Sicily, which became known as Magna Graecia, or Great Greece.

The Ionian cities eventually lost their freedom, first to the Lydians and then to the Persians, whose attempt to conquer Greece ended with

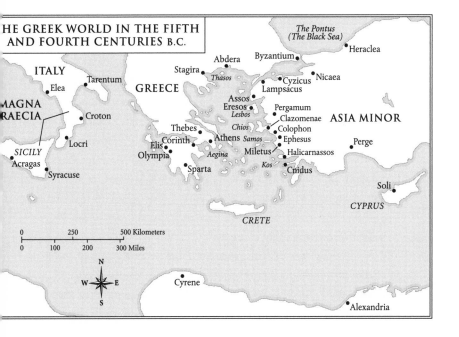

their defeat by the Greek allies at the battle of Plataea in 479 B.C. The Persian king Xerxes took his revenge for this defeat by destroying Miletus, but the city was soon afterward rebuilt and by the middle of the fifth century B.C. it was once again a flourishing port and commercial center.

The far-ranging maritime activities of the Milesians brought them into contact with older and more advanced civilizations in the Middle East, particularly in Egypt, from which the Greeks returned with ideas as well as goods. Herodotus writes that "the Egyptians by their study of astronomy discovered the solar year and were the first to divide it into twelve parts—and in my opinion their method of calculation is better than the Greek."

The trade routes of the Milesians also took them to Mesopotamia, where they probably acquired the knowledge of astronomy they needed for celestial navigation and timekeeping. They obtained the gnomon, or shadow marker, in Mesopotamia, according to Herodotus, who says that "knowledge of the sundial and the gnomon and the twelve divisions of the day came into Greece from Babylon." The gnomon was also used to determine the equinoxes, when the sun rises due east and sets due west, as well as the winter and summer solstices, when the noon shadow is longest and shortest, respectively.

The Greek word for star, *aster,* is derived from Ishtar, the Babylonian fertility goddess, whom the Greeks identified with the planet Venus. They at first thought that Venus was two different stars, calling it Eosphoros when it was seen before sunrise and Hesperos when it appeared in the evening. They later realized that the morning and evening stars were the same celestial body, which they called Aphrodite, the goddess of love, thus perpetuating the cult of Babylonian Ishtar. Venus is the only planet mentioned by Homer, who in the *Iliad* calls it Eosphorus when describing the funeral of Patroklos, and Hesperos when telling of the battle between Achilles and Hektor. Sappho also sings of Venus alone among the planets, and then only as Hesperos, "fairest of all the stars that shine."

The Ionian Greeks soon progressed far beyond their predecessors intellectually, particularly in Miletus, which in the last quarter of the sixth century B.C. gave birth to the first three philosophers of nature. All that is known of their thought are fragmentary quotes or paraphrases of

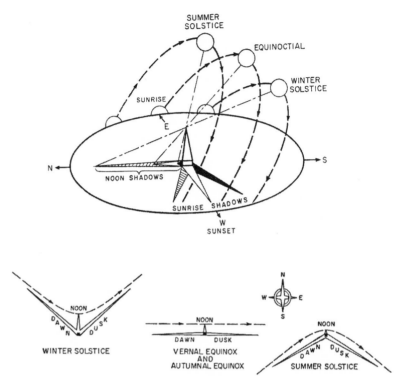

The use of a gnomon in determining the seasons. The examples here are from the middle northern latitudes. The upper drawing shows the seasonal variations of the sun's path and the shadows that it casts at noon and sunset; the term "equinoctial" (equal day and night) refers to both the vernal (spring) and autumnal (fall) equinoxes. The lower drawings show the gnomon's shadows at the solstices and equinoxes. (*from Kuhn, 1957*)

their works by later writers. Aristotle referred to them as *physikoi*, or physicists, from the Greek *physis*, meaning "nature" in its widest sense, contrasting them with the earlier *theologoi*, or theologians, for they were the first who tried to explain phenomena on natural rather than supernatural grounds. Earthquakes, for example, which both Homer and Hesiod attributed to the action of Poseidon, the "earth shaker," were explained by Thales as the rocking of the earth while it floated in the all-encompassing waters of Oceanus.

Plato listed Thales (ca. 625–ca. 547 B.C.) among the Seven Sages of ancient Greece, while Aristotle considered him to be the "first founder" of Ionian natural philosophy. There is a tradition that Thales visited Egypt, where he is supposed to have calculated the height of a pyramid by pacing off its shadow, doing this at the time of day when the height of any object is equal to the length of its shadow. Herodotus credits Thales with having predicted the total eclipse of the sun visible in central Asia Minor on 28 May 585 B.C., when the Lydians and Persians were at war. Given the state of knowledge at the time, it would have been impossible for Thales to predict that an eclipse might be visible in that region, but once he became enshrined as one of the Seven Sages all sorts of intellectual accomplishments were attributed to him, including the first geometrical theorems known to the Greeks.

The most enduring ideas of the Milesian physicists proved to be their speculations on the nature of matter, particularly their belief that there was an *arche*, or fundamental substance, that endured through all apparent change. Aristotle writes that "Thales, who led the way in this kind of philosophy, says that the principle is water, and for this reason declared that the earth rests on water."

Aristotle thought that Thales chose water as the *arche* "from the observation that the nourishment of all creatures is moist . . . and water is for most moist things the origin of their nature." His choice of water was undoubtedly because it is normally a liquid but when heated becomes a vapor and when frozen is transformed to solid ice, so that the same substance appears in all three forms of matter. More fundamentally, Thales was trying to answer a question that marks the beginning of Greek philosophy: What is the nature of the reality behind phenomena?

Anaximander (ca. 610–ca. 545 B.C.) was a younger friend of Thales's and a fellow citizen of Miletus. Following the tradition that Thales left no writings, Themistius (ca. 317–ca. 388) describes Anaximander as "the first of the Greeks, to our knowledge, who was bold enough to publish a treatise on nature." Ancient sources also attribute to Anaximander books on astronomy, where he is said to have used the gnomon to determine "solstices, times, seasons and equinoxes," as well as a work on geography in which he was the first to draw a map of the *ecumenos*, the inhabited world.

Anaximander called the fundamental substance *apeiron*, the "boundless"; the term is sometimes translated as "the infinite," meaning that it is

to the stream of a river he says you would not step twice into the same river."

The relative stability of nature was the result of what Heraclitus called the opposite tension, a balance of opposing forces producing equilibrium, and the unity of the cosmos was due to *Logos*, or Reason, which gives order to the natural world. He held that divinity was the unity of opposites, as in his statement that "God is day night, winter summer, war peace, satiety hunger, he undergoes alteration in the way that a fire, when it is mixed with spices, is named according to the scent of each of them."

Heraclitus believed that the evidence of the senses is deceptive and must be used with caution, since it deals with transitory phenomena, as he says in one of his aphorisms: "Evil witnesses are eyes and ears for men, if they have souls that do not understand their language."

As science developed, *physikoi* extended one branch or another of what had already been begun. Hecataeus of Miletus (fl. ca. 500 B.C.), a contemporary of Heraclitus's, is credited with following the lead of Anaximander in drawing a map of the world known to the Greeks. As a supplement to his map he also wrote a work entitled *Periegesis*, a "guide" or "journey round the world," a description of the countries and people to be seen on a coastal voyage around the Mediterranean and the Black Sea, along with some excursions inland, ranging as far as Scythia, Persia, and India. The enormous extent of his map is a measure of how far abroad the Greeks had traveled in their colonization and trade, exposing them to cultures around the Mediterranean and the Black Sea.

During the third quarter of the sixth century B.C. the Ionian intellectual enlightenment spread to Magna Graecia, sparked by two of the most original minds of the archaic period, Pythagoras and Xenophanes.

Pythagoras (ca. 560–ca. 480 B.C.) was born on Samos, one of the two Aegean islands that were part of the Panionic League, lying off the Aegean coast of Asia Minor northwest of Miletus. There is a late tradition that in his youth Pythagoras traveled to both Egypt and Babylonia to study mathematics. When Pythagoras came of age he fled from the rule of Polycrates, the tyrant of Samos, and moved to Croton in southern Italy, a Greek colony founded in the eighth century B.C. There he established a society that was both a scientific school and a religious sect; its beliefs included that of metempsychosis, or the transmigration of souls. Otherwise little is known of Pythagoras himself, and it is impossible to distinguish his own ideas from those of his followers.

The Pythagoreans are credited with laying the foundations of Greek mathematics, particularly geometry and the theory of numbers. The most famous of their discoveries is the Pythagorean theorem, which says that in a right triangle the square of the hypotenuse equals the sum of the squares of the other two sides.

Their religious beliefs also led the Pythagoreans into numerology, or number mysticism, including such notions that odd numbers have male characteristics and even ones female. The most sacred number is ten, the sum of the first four numbers, where one is the "atom" of numbers, of which two generate a line, three if not all in a line determine a plane, and four if not all in the same plane form the vertices of a solid. Arrayed one above another as a series of dots, or "figurative numbers"—that is, 1(.), 2(..), 3(...), 4(....)—the first four integers form an equilateral figure known as the *tetractys,* the number of the universe, since it is the sum of all possible dimensions. The *tetractys* became the symbol of the Pythagoreans, who in time acquired the reputation of being magicians and sorcerers. The third-century church father Hippolytus writes in his *Philosophumena* of "magical arts and Pythagorean numbers" and notes that "Pythagoras also touched on magic, as they say, and himself discovered an art of physiognomy, laying down as a basis certain numbers and measures."

Another idea attributed to the Pythagoreans is the concept of "cosmos"—*kosmos* in Greek—which a modern dictionary defines as an "orderly, harmonious and systematic universe." The original Greek meaning of *kosmos* is given by Plato in a passage from *Meno,* where he appears to be referring to the Pythagoreans. "The wise men," he says, "tell us that heaven and earth, gods and men are bound together by kinship and love and orderliness and temperance and justice; and for this reason, my friend, they give to the whole the name of *kosmos,* not a name implying disorder or licentiousness."

According to tradition, the Pythagoreans were the first to recognize the numerical relations involved in musical harmony, to which they were led by their experiments with stringed instruments. This made them believe that the cosmos was designed by a divine intelligence according to harmonious principles, and that this harmony could be expressed in terms of numbers. As Aristotle wrote of the Pythagoreans: "They supposed the elements of numbers to be the elements of all things, and the whole heaven to be a musical scale and a number."

Philolaus of Crotona, who flourished in the second half of the fifth century B.C., is supposed to have written a comprehensive work on Pythagorean cosmology. According to Philolaus, the Pythagoreans believed that the earth was not stationary but moved in a circle around a central fire called Hestia, the hearth of the cosmos, along with the sun, the moon, the stars, and the five visible planets—Mercury, Venus, Mars, Jupiter, and Saturn—as well as another body known as the "counter-earth," invisible because it was on the opposite side of the cosmos. They thought that the celestial bodies moved in such a way as to create a celestial harmony, as Alexander of Aphrodisias writes in his commentary on Aristotle's *Metaphysics*: "They said that the bodies which revolve around the center have their distances in proportion, and some revolve more quickly, others more slowly, the sounds which they make being deep in the case of the slower, and high in the case of the quicker; these sounds then, depending on the ratio of the distances, are such that their combined effect is harmonious."

Aristotle writes of how the Pythagoreans explained why we do not hear this celestial harmony: "They account for it by saying that the sound is with us right from birth and has thus no contrasting silence to show it up; for voice and silence are perceived by contrast with one another, and so all mankind is undergoing an experience like that of a coppersmith, who becomes by long habit indifferent to the din around him." Thus ordinary mortals are not aware of the divine harmony, as Lorenzo explains to Jessica in *The Merchant of Venice*.

> There's not the smallest orb which thou behold'st
> But in his motion like an angel sings.
> Still quiring to the young-eyed cherubims;
> Such harmony is in immortal souls,
> But, whilst this muddy vesture of decay
> Doth grossly close it in, we cannot hear it.

The poet and philosopher Xenophanes (ca. 570–after 478 B.C.), a somewhat older contemporary of Pythagoras's, was born in the Ionian city of Colophon, northwest of Ephesus, and is said to have been a disciple of Anaximander's. He fled from Ionia after it was conquered by the Persians in 545 B.C. and moved to Magna Graecia, where, according to Diogenes Laertius, he lived in Sicily at Zancle and Catana, two Greek colonies founded in the eighth century B.C. He became the poet of the Ionian enlightenment in the West, and though his extant fragments are primarily of literary rather than scientific interest, some of his seminal ideas profoundly influenced the development of natural philosophy in Magna Graecia.

Xenophanes objected to the anthropomorphic polytheism of Homer and Hesiod, whom he condemned for having "ascribed to the gods all deeds that among men are a shame and a reproach: thieving, adultery and mutual deception." He said that men make gods in the image of themselves, so that "the Ethiopians say their gods are snub-nosed and black, the Thracians that theirs have light blue eyes and red hair."

His own view was both monotheistic and pantheistic, as is evident from one of his fragments: "God is one, greatest among gods and men, in no way like mortals either in body or mind. He sees as a whole, perceives as a whole, hears as a whole. Always he remains in the same place, not moving at all, nor does it befit him to go here and there at different times; but without toil he makes all things shiver by the impulse of his mind."

The Pythagorean notion of the transmigration of souls was ridiculed by Xenophanes, who in one of his poems tells the story of how Pythagoras stopped a man beating a dog, saying to him "Stop, do not beat him. It is the soul of a friend, I recognize his voice!"

Tradition has it that Xenophanes was the teacher of Parmenides (ca. 515–ca. 450 B.C.) of Elea, a colony founded on the Tyrrhenian coast of Italy by the Ionian city of Phocaea. Like Heraclitus, Parmenides stressed the unreliability of the senses, saying that one should not "let custom, born of much experience, force thee to let wander along this road thy aimless eye, thy echoing ear or thy tongue, but do thou judge by reason the strife-encountered proof that I have spoken."

Whereas Heraclitus believed that everything was in a state of flux and nothing was permanent, Parmenides denied absolutely the possibility of motion and any other kind of change, holding that they were mere illusions of the senses. As he wrote in his didactic poem *The Way of Truth*,

"Either a thing is or it is not," meaning that creation or destruction or any other kind of change was impossible, including motion.

Parmenides did not admit the existence of multiplicity and time; all that exists, he said, is one and now. His cosmos is a full, uncreated, indestructible, changeless, motionless, eternal, and perfect sphere of being, and all sensory evidence to the contrary is illusion. Echoes of this immutable Parmenidean cosmology reverberated from antiquity down to the European Renaissance, as in the last canto of Spenser's *The Faerie Queene:*

> *Then gin I thinke on that which Nature sayd,*
> *Of that same time when no more Change shall be,*
> *But stedfast rest of all things, firmely stayd*
> *Upon the pillours of Eternity,*
> *That is contrayr to Mutabilitie;*
> *For all that moveth doth in Change delight.*

The philosophy of Parmenides was vigorously defended by his pupil Zeno of Elea (ca. 490–ca. 425 B.C.), who proposed several paradoxes designed to show that apparent motion is illusory. One of these concerns a hypothetical race between Achilles and a tortoise, which is given a head start to make up for its slowness. Achilles runs to catch up, but to do so he must first reach the point from which the tortoise started, by which time it will have moved farther, and likewise in each diminishing interval that follows. The number of such intervals is infinite, according to Zeno, and although the times become increasingly shorter their sum is limitless. Thus Achilles will never catch the tortoise, and the conclusion is that motion is an illusion of the senses. This and the other paradoxes of Zeno were not fully resolved until the second half of the nineteenth century, when mathematicians proved that the sum of an infinite series, such as that involved in the race between Achilles and the tortoise, can be finite.

Some of the profound questions raised by Parmenides were addressed by Empedocles (ca. 482–ca. 432 B.C.) of Acragas, another Greek colony in Sicily. Empedocles was the author of two hexameter poems, one entitled *On Nature* and the other *Purifications*, of which a total of 450 lines have survived in the form of quotations by Aristotle and other later writers. While Empedocles agreed with Parmenides that there was a serious problem regarding the reliability of sense impressions, he insisted that we are utterly dependent on our senses, for they are our only direct contact with the world of nature. He warned, however, that we must carefully evaluate the information obtained by our senses to gain true knowledge. "But come, consider with all thy powers how each thing is manifest, neither holding sight in greater trust as compared with hearing, nor loud-sounding hearing above the clear evidence of thy tongue, nor withhold thy trust from any of the other limbs [organs], wheresoever there is a path for understanding, but think on each thing in the way in which it is manifest."

According to Aristotle, Empedocles was the first to say that there were four fundamental substances—earth, air, fire, and water—which he called the "roots of everything." Referring to the four elements, he said that "from these things sprang all things that were and are and shall be, trees and men and women, beasts and birds and water-bred fishes, and the long-lived gods too, most mighty in their prerogatives." He has the four substances intermingling and separating under the influence of two forces that he calls Love and Strife. He says "the elements are continually subject to an alternate change, at one time mixed together by Love, at another separated by Strife."

Empedocles thus introduced the concept of force as distinct from matter as the physical cause for the phenomena in nature. He pictured the cosmos as existing in a state of dynamic equilibrium between the opposing forces, with motion taking place when either predominates. His identification of earth, water, and air as elements corresponds to the modern classification of matter into solid, liquid, and gaseous states. Fire to him represented not only flames but phenomena occurring in the heavens, such as lightning and comets. The Empedoclean theory of the four elements was one of the most enduring in the history of science, lasting for more than two thousand years. It left its impress on literature as well as science, as evidenced by the lines in *The Faerie Queene*, where Spenser writes of how the four elements "the groundwork bee / Of all the world and of living wights." He continues:

To thousand sorts of Change we subject see:
Yet are they chang'd (by other wondrous slights)

Into themselves, and lose their native mights;
The Fire to Ayre, and th' Ayre to Water sheere,
And Water into Earth; yet Water fights
With Fire, and Ayre with Earth, approaching neere:
Yet all are in one body, and as one appeare.

Empedocles had several other original ideas. He held that light travels through space with great but finite speed. He was also the first to show that air, although invisible, is a real physical substance. He demonstrated this by taking a vessel called a clepsydra, or water clock, and immersing it upside down, showing that no liquid entered until the air was allowed to escape from inside the vessel in equal quantity.

Some of the statements made by Empedocles gave rise to legends that he was a divine healer and a wonder worker. "I, an immortal god," he says in one of his fragments, "no longer a mortal, go about among you all, honored as is fitting, crowned with fillets and flowery garlands." He goes on to describe how his disciples crowded about him, "asking the path to gain, some desiring oracles, while others seek to hear the word of healing for all kinds of diseases." One of the stories told about Empedocles was that he left the world by jumping into the volcanic crater of Mount Etna, leaving only his sandals behind, though other legends have him ending his days as an exile in the Peloponnesos.

A radically different theory of matter from those of Empedocles and Parmenides was later proposed by Leucippus, who was probably born in Miletus late in the sixth century B.C. and moved to Abdera in Thrace, which had been founded circa 500 B.C. by refugees from the Ionian city of Teos. His lost work *The Greater World System* apparently originated the atomic theory, which is usually credited to his pupil Democritus.

Democritus (ca. 470–ca. 404 B.C.) was born at Abdera and is reported to have visited Athens, but no one knew him there, according to the Greek biographer Diogenes Laertius. His version of the atomic theory appeared in a book entitled the *Little World System*, which he may have so called out of deference to the work of his teacher Leucippus.

The atomic theory of Leucippus and Democritus holds that the *arche* exists in the form of atoms, the irreducible minima of all physical substances, which through their ceaseless motion and mutual collisions take on all of the various forms observed in nature. The only extant fragment by Leucippus himself says, "Nothing occurs at random, but everything for a reason and by necessity." By this he meant that atomic motion is not chaotic but obeys the immutable laws of nature.

According to Democritus, there is no limit either to the number of atoms or the extent of the void, and so innumerable worlds are possible, of which our cosmos is only one. One of his surviving fragments quotes Democritus as saying that

> there are innumerable worlds of different sizes. In some there is neither sun nor moon, in others they are larger than in ours and others have more than one. These worlds are at irregular distances, more in one direction and less in another, and some are flourishing, others declining. Here they come into being, there they die, and they are destroyed by collision with one another. Some of the worlds have no animal or vegetable life nor any water.

Another Democritian fragment has him saying that he was younger but contemporary with the philosopher Anaxagoras, who was born circa 500 B.C. in the Ionian city of Clazomenae, which he left for Athens at the age of twenty. Anaxagoras was the first philosopher to live in Athens; he resided there for thirty years, becoming a teacher and close friend of Pericles'.

The views of Anaxagoras on the nature of matter were even more pluralistic than those of Empedocles, for he postulated the existence of a very large number of elements in his "seed theory." "We must suppose," he writes, "that there are many things of all sorts in things that are aggregated, seeds of all things, with all sorts of shapes and colors and tastes. . . . There is a portion of everything in everything." He also postulated an element called the *aether,* which was in constant rotation and carried with it the celestial bodies. He said that "the sun, the moon and all the stars are red-hot stones which the rotation of the *aether* carries round with it." The *aether* proved to be a very enduring concept, and it reappeared in cosmological theories up until the early twentieth century.

Another of his ideas concerned what the Greeks of his time called *Nous,* or Mind, by which he meant the directing intelligence of the cosmos, as opposed to inert matter. This earned for Anaxagoras the nickname of Nous, as Plutarch notes in his *Life of Pericles*:

> But the man who most consorted with Pericles ... was Anaxagoras the Clazomenian, whom men of those days used to call "*Nous*" either because they admired that comprehension of his, which proved of such surpassing greatness in the investigation of nature; or because he was the first to enthrone in the universe not Chance, nor yet Necessity, but Mind (*Nous*) pure and simple, which distinguishes and sets apart, in the midst of an otherwise chaotic mass, the substances which have like elements.

Around 450 B.C. the enemies of Pericles indicted Anaxagoras on charges of impiety and "medeism"—being pro-Persian. Aided by Pericles, Anaxagoras was able to escape to Lampsacus on the Hellespont, where he founded a school that he directed for the rest of his days. After the death of Anaxagoras, circa 428 B.C., the people of Lampsacus erected a monument to his memory in their agora, the market quarter, dedicating it to mind and truth, which were at the core of his philosophy. The anniversary of his death was for long afterward commemorated in Lampsacus, and by his dying request the students of the city were always let out of school on that day.

Anaxagoras was the last of the Ionian physicists, for even in his own lifetime Athens had replaced Ionia as the common meeting place for philosophers of nature. Xenophanes attributed the decline of Ionia to the corrupting wealth of its citizens, as he writes in one of his poems.

> *And they learned dainty, useless Lydian ways*
> *While they were still from hated tyrants free.*
> *In robes all scarlet to the assembly went*
> *A thousand men, no less: vainglorious,*
> *Preening themselves on their fair flowing locks,*
> *Dripping with scent of artificial oils.*

Such was the world of Ionia, where the first physicists began to speculate about the nature of the cosmos and the bounds of knowledge.

Their immediate successors brought philosophy to Magna Graecia and Athens, the first stages in a journey that would take scientific ideas and theories back and forth between East and West like birds on the wing, continuing their flight long after Miletus and the other Ionian cities were reduced to utter ruins.

CLASSICAL ATHENS: THE SCHOOL OF HELLAS

The ruins of ancient Athens are still at the heart of the modern city, crowned by the Parthenon, the magnificent temple of Athena built in the mid-fifth century B.C. by Pericles. As Thucydides says, quoting Pericles' paean to the greatness of Athens: "Mighty indeed are the marks and monuments of our empire which we have left. Future ages will wonder at us, as the present age wonders at us now."

A short stretch of the ancient walls of Athens, built by Themistocles in 478 B.C., can still be seen in the Theseion quarter of the modern city. They are within the archaeological site of the ancient Kerameikos cemetery, which is just outside the two main gates in the Themistoclean walls, the Dipylon Gate and the Sacred Gate. The latter took its name from the Sacred Way, the route of the processions that led from Athens to the great shrine of Eleusis, while the Dipylon was the beginning of the road known as the Dromos.

From the sixth century B.C. onward many of the leading figures in Athenian history were buried along the sides of these two roads, which were part of the Demosion Sima, the state burial ground. This was where Pericles delivered his famous funeral oration in 431 B.C., honoring the Athenians who fell in the first year of the Peloponnesian War. He reminded his fellow citizens that they were fighting to defend a free and democratic society that was "open to the world," one whose "love of the things of the mind" had made their city "the school of Hellas."

The course of the ancient Dromos is today the route of Odos Platonos, which leads from the Kerameikos cemetery to the quarter known as Academia, a distance of about one Attic mile (some 1,200 paces) outside the walls of ancient Athens. This quiet residential area takes its name from the famous Platonic Academy, whose site has been partially excavated, though there is very little left to see of the buildings of what for more than nine centuries was the most renowned school of Hellas.

The Academy was named for an ancient shrine of Hekademos, an earth-born hero of Attic mythology, who is supposed to have planted here twelve olive trees that were cuttings from Athena's sacred olive on the Acropolis, her gift to the people of Attica. The *temenos*, or sacred enclosure of the shrine, was vast, judging from the extent of the excavations, with a periphery of about half a mile. Plutarch says that the grounds were first enclosed and developed by Cimon, who transformed it "from a waterless and arid spot into a well-watered grove, with clear running tracks and shady groves." There was already a gymnasium here by the time of Aristophanes, for in *The Clouds,* produced in 423 B.C., one of the characters describes the footraces that took place within the groves of academe:

> You will spend your time, sleek and blooming, in the gymnasiums. . . . You will go down to the Academy and run races under the sacred olives with a virtuous comrade, crowned with white reeds and smelling of bindweed and careless ease and the white poplar that sheds its leaves, happy in the springtide when the plane-tree whispers to the elm.

Plato (427–347 B.C.) was born two years after the death of Pericles. He was profoundly influenced by Socrates, of whom he writes in many of his dialogues. In his dialogue *Phaedo,* or *On the Soul,* Plato describes the last hours of Socrates before he was forced to commit suicide in 399 B.C. in the state prison, having been convicted of corrupting the youth of Athens through his subversive ideas.

After the death of Socrates, Plato left Athens and traveled abroad, visiting Italy and Sicily. He returned to Athens in 386 B.C. and a few years later founded the Academy. Other schools and institutions functioned within and around the *temenos* of Hekademos, but in time the gymnasium founded by Plato became so famous that the name Academy came to be applied to it alone. Milton describes it in *Paradise Regained* as "the olive grove of Academe, Plato's retirement, where the Attic bird trills her thick-warbl'd notes the summer long."

Virtually nothing is known of the school's formal organization or its curriculum. At least in its early years it may have been patterned on the educational system described by Plato in his *Republic* and *Laws*, particularly in Book I of the latter, where he writes that "what we have in mind is education from childhood in virtue, a training which produces a keen desire to become a perfect citizen who knows how to rule and be ruled as justice demands."

The Academy probably corresponded to the colleges of the first European universities, with a community of scholars sharing a common table. Athenaeus of Naucratis (fl. ca. A.D. 200) writes that "the philosophers make it their business to join with their students in feasting according to certain set rules." Plato, in the *Laws*, says that symposia were held according to the rules of a master of ceremonies, who must himself remain sober. Antigonus of Carystus (fl. 240 B.C.) writes that Plato did not hold these symposia just for the sake of carousing till dawn, "but that they might manifestly honor the gods and enjoy each other's companionship, and chiefly to refresh themselves with learned discussion."

Plato's dialogues also mention some of the other philosophers who were in Athens in the time of Socrates (469–399 B.C.) and in his own era. The *Parmenides* is based on a supposed visit that Parmenides made to Athens in his old age, when he and his follower Zeno met the young Socrates. Plato writes that "Zeno and Parmenides once came to the Great Panathenaea. Parmenides was already quite venerable, very gray but of distinguished appearance, about sixty-five years. Zeno was at the time close to forty ... Socrates was then quite young."

Plato's dialogue *Protagoras* mentions a young man who goes to the Aegean island of Kos to study medicine under Hippocrates the Asclepiad. The renowned physician Hippocrates (460–ca. 370 B.C.) of Kos was an older contemporary of Plato's. He was known as the Asclepiad because he belonged to one of the families that perpetuated the cult of Asclepios, the god of healing, whose first shrines were founded circa

500 B.C. The most famous of these healing shrines were the Asclepieia at Epidaurus, Athens, and Pergamum, besides which there were also renowned medical schools at Kos and Cnidus.

The writings of Hippocrates and his followers, the so-called Hippocratic Corpus, comprises some seventy works dating from his time to circa 300 B.C. Besides treatises on the various branches of medicine, they include clinical records and notes of lectures given to the general public on medical topics. One of the treatises, on deontology, or medical ethics, contains the famous Hippocratic oath, which is still taken by physicians today. One work in the Hippocratic Corpus is entitled *The Sacred Disease*, for the name given to epilepsy; those suffering from it were believed to be stricken by the gods. The author of this work, who may be Hippocrates himself, says that epilepsy, like all other diseases, has a natural cause, and that those who first called it sacred were merely trying to cover up their ignorance.

Plato's attitude toward the study of nature is evident from what he has Socrates say in the *Phaedo*. There Socrates tells of how he had been attracted to the ideas of Anaxagoras because of his concept of *Nous*, or Mind. But he was ultimately disappointed, for he found that Anaxagoras did not use Mind to explain the element of design or order in nature, giving materialistic reasons instead. "This magnificent hope was dashed as I went on reading," he says, "and saw that the man made no use of Mind, nor gave it any responsibility for the management of things, but mentioned as causes air and aether and water and many other strange things."

Socrates was disillusioned by Anaxagoras and the other early natural philosophers because they only told him *how* things happened rather than *why*. What Socrates was looking for was a teleological explanation, one involving evidences of design in nature, for he believed that everything in the cosmos was directed toward attaining the best possible end. Plato's own ideas in science are contained principally in the *Timaeus*, where he presents a cosmology that he says is "only along the lines of the likely stories we have been following." Nevertheless, the *Timaeus* was enormously influential down to the time of the European Renaissance.

Plato's attitude toward astrology is revealed in the *Timaeus*, where he writes of the "everlasting and unwandering stars—divine, living things," an expression echoed in medieval astrological writings. And in the

Republic he speaks of the harmony of the heavenly spheres and of "the spindle of Necessity, by means of which all the revolutions are turned," suggesting that the human soul is subject to the motions of the celestial bodies.

Over the entrance of the Academy there was said to have been an inscription stating "Let no one ignorant of geometry enter here." This probably derives from Plato's *Republic,* where Socrates says that "we must require those in our fine city not to neglect geometry in any way, for even its by-products are not insignificant."

Plato believed that mathematics was a prerequisite for the dialectical process that would give future leaders the philosophical insight necessary for governing a state. The mathematical study included arithmetic, plane and solid geometry, harmonics, and astronomy. Harmonics involved a study of the physics of sound as well as an analysis of the mathematical relations supposedly developed by the Pythagoreans in their researches on music. Astronomy was studied not only for its practical applications, but for what it revealed of the "true numbers" and "true motions" behind the apparent movements of the celestial bodies.

Plato's most enduring influence on science was his advice to approach the study of nature, particularly astronomy, as an exercise in geometry. Through this "geometrization of nature," applicable only in those disciplines such as mathematical astronomy that could be suitably idealized, one can arrive at relations that were as "certain" as those in geometry. As Socrates remarks in the *Republic:* "Let's study astronomy by means of problems, as we do geometry, and leave the things in the sky alone."

The principal problem in Greek astronomy was to explain the motion of the celestial bodies—the stars, sun, moon, and the five visible planets. As seen from the earth, the celestial bodies all seem to rotate daily about a point in the heavens called the celestial pole, actually the projection of the earth's north pole among the stars. This apparent motion is actually due to the axial rotation of the earth in the opposite sense. Although the sun rises in the east and sets in the west, each day its position among the stars as it rises appears to be about one degree back toward the west, making the transit of the twelve signs of the zodiac in one year, an apparent motion produced by the orbiting of the earth around the sun.

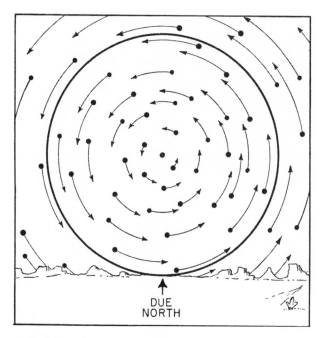

The apparent motion of typical stars in the northern sky over a two-hour period, where the center of rotation is the north celestial pole, the projection of the north geographic pole. (*from Kuhn, 1957*)

The apparent path of the sun through the zodiac, the so-called ecliptic, makes an angle of about 23.25 degrees with the celestial equator, the projection of the earth's equator among the stars. This is due to the fact that the earth's axis is tilted by 23.25 degrees with respect to the perpendicular of the ecliptic plane, an obliquity that is responsible for the recurring cycle of seasons. The obliquity of the ecliptic actually varies cyclically between 22.1 and 24.5 degrees over a period of about forty thousand years, and in the classical Greek era it was about 23.5 degrees.

The planets all follow paths that are close to the ecliptic, moving from east to west during the night along with the fixed stars, while from one night to the next they generally move slowly back from west to east around the zodiac. Each of the planets also exhibits a periodic retrograde motion, which shows as a loop when its path is plotted on the celestial sphere. This is due to the fact that the earth is moving in orbit around the sun, passing the slower outer planets and being itself passed by the swifter inner planets, the effect in both cases making it appear that the planet is moving backward for a time among the stars.

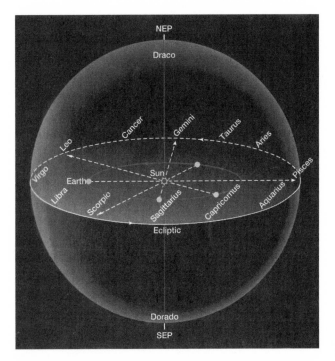

The apparent motion of the sun among the constellations; the effect is due to the fact that the observer, on the earth, is orbiting around the sun.

According to Simplicius (ca. 490–ca. 560), Plato posed a problem for those studying the heavens: to demonstrate "on what hypotheses the phenomena [i.e. the "appearances," in this case the apparent retrograde motions] concerning the planets could be accounted for by uniform and ordered circular motions."

The first solution to the problem was provided by Eudoxus of Cnidus (ca. 400–ca. 347 B.C.), a younger contemporary of Plato's at the Academy. Eudoxus was the greatest mathematician of the classical period, credited with some of the theorems that would later appear in the works of Euclid and Archimedes. He was also the leading astronomer of his era and had made careful observations of the celestial bodies from his observatory at Cnidus, on the southwestern coast of Asia Minor. (An observatory at that time would have comprised little more than a few simple instruments for sighting on the celestial bodies and determining their positions in the heavens.) Eudoxus suggested that the path of each of the five planets was the result of the uniform motion of four connected spheres, all of which had the earth as their center, but with their

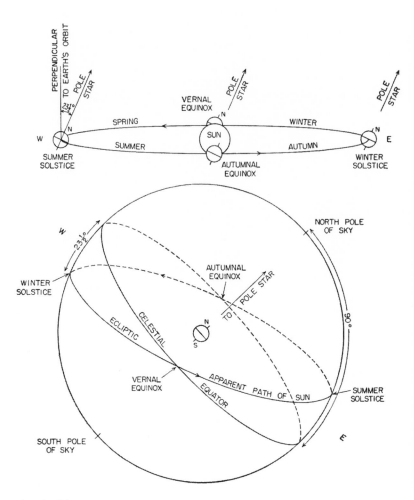

The tilt of the earth's axis as the cause of the seasons.

axes inclined to one another and rotating at different speeds, the planet being attached to the equator of the innermost sphere and the outermost one moving with the fixed stars. The motions of the sun and the moon were accounted for by three spheres each, while a single sphere sufficed for the daily rotation of the fixed stars, making a total of twenty-seven spheres for the cosmos. Eudoxus's model, known as the theory of homocentric spheres, was elaborated upon by Callipus of Cyzicus (fl. 370 B.C.), who added two spheres each for the sun and moon, as well as one each for Mercury, Venus, and Mars, to make a total of thirty-four.

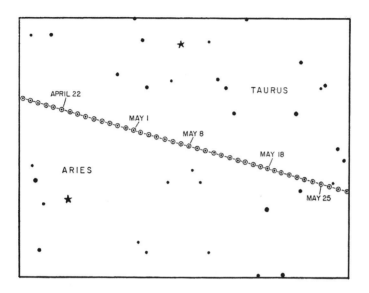

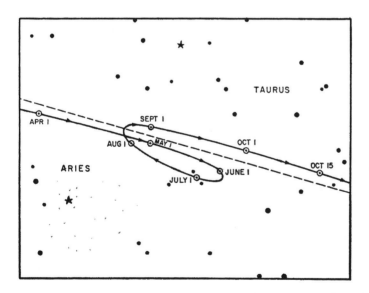

Above: The apparent motion of the sun through the constellations Aries and Taurus.
Below: The apparent motion of Mars through the constellations Aries and Taurus, showing its retrograde motion. (*from Kuhn, 1957*)

The theory of homocentric spheres was subsequently adopted by Aristotle as the physical model for his geocentric cosmos, using fifty-five planetary spheres plus another for the fixed stars.

Aristotle (384–322 B.C.) was born at Stagira in Macedonia. His father, Nicomachus, served as physician to the Macedonian king Amyntas III, in whose court Aristotle received his early education. At the age of seventeen he moved to Athens, in order to enroll in Plato's Academy, where he remained for twenty years. After the death of Plato in 347 B.C. Aristotle moved to Assos, on the northwestern coast of Asia Minor, where he entered the service of the tyrant Hermeias. Hermeias had been a student of Plato's and sought to create at Assos the ideal state described in the *Republic,* inviting Aristotle and other scholars to teach there, including Theophrastus of Eresos in Lesbos.

Aristotle remained in Assos until 344 B.C., when Hermeias was executed by the Persians. He then moved across to the island of Lesbos, where he and Theophrastus continued the pioneering studies in botany they had begun at Assos. After a year there Aristotle left for the Macedonian capital of Pella to enter the service of Philip II, serving as tutor to the king's son and eventual successor, Alexander the Great.

Aristotle returned to Athens in 335 B.C., the year after Alexander succeeded to the Macedonian throne. That same year he founded a gymnasium called the Lyceum, which would rival the Academy in its fame. Aristotle continued to teach and do research in the Lyceum until 323 B.C., when the death of Alexander was followed by an anti-Macedonian movement that forced him to leave Athens and return to Macedonia; he died there the following year.

Aristotle's writings are encyclopedic in scope, including works on logic, metaphysics, rhetoric, theology, politics, economics, literature, ethics, psychology, physics, mechanics, astronomy, meteorology, cosmology, biology, botany, natural history, and zoology. Thus Montaigne was led to write of "Aristotle that hath an oare in every water, and medleth with all things."

The dominant concept in Aristotle's philosophy of nature is the principle of teleology, the idea that natural processes are directed toward an end. This is stated most clearly in the second book of his *Physics:* "Now intelligent action is for the sake of an end; therefore the nature of things also is so: and as in nature. Thus if a house, e.g., had been a thing made by nature, it would have been made in the same way as it is now by art; and if things made by nature were made also by art, they would come to be in the same way as by nature."

The main outlines of Aristotle's theory of matter and his cosmology derive from earlier Greek thought, which distinguished between the imperfect and transitory terrestrial world below the sphere of the moon and the perfect and eternal celestial region above. He took from the Milesian physicists the notion that there was one fundamental substance in nature, and reconciled this with Empedocles' concept of the four terrestrial elements—earth, water, air, and fire—to which he added the *aether* of Anaxagoras as the basic substance of the celestial region.

According to Aristotle, the fundamental terrestrial substance, which he called *prostyle,* is completely undifferentiated. It has no qualities whatsoever; that is, it has no definite size, shape, place, weight, color, taste, smell, or the like, being in effect the utterly characterless raw material out of which the world is made. When this matter takes on various qualities it becomes one of the four terrestrial elements, and through further developments it takes on the form of the things seen in the world. Aristotle would describe this as matter taking on form. The matter is the raw material, the form is the collection of all the qualities that give an object its distinctive character. These two aspects of existence—matter and form—are inseparable, and can exist only in conjunction with one another.

Aristotle assigned to each of the four terrestrial elements two qualities, one from each of the two pairs of opposites: hot-cold and dry-moist. Thus earth was dry and cold, water cold and moist, air moist and hot, fire hot and dry. These elements were not immutable; any one of them could be transformed into any other one if either or both of its basic properties changed into its opposite.

Aristotle's cosmology arranged the four elements in order of density, with the immobile spherical earth at the center surrounded by concentric shells of water (the ocean), air (the atmosphere), and fire, which included not only flames but extraterrestrial phenomena such as lightning, rainbows, and comets. The natural motion of the terrestrial elements was to their natural place, so that if earth is displaced upward in air and released it will fall straight down, whereas air in water will rise, as does fire in air. This linear motion of the terrestrial elements is tempo-

Aristotle's cosmology.

rary since it ceases when they reach their natural place. Aristotle's theory of motion has heavier objects falling faster than those that are light, one of two of his erroneous ideas that dominated physics until the seventeenth century, the other being the impossibility of a void.

According to Aristotle, the celestial region begins at the moon, beyond which are the sun, the five planets, and the fixed stars, all embedded in crystalline spheres rotating around the immobile earth. The celestial bodies are made of *aether*, the quintessential element, whose natural motion is circular at constant velocity, so that the motions of the celestial bodies, unlike those of the terrestrial region, are unchanging and eternal.

Heraclides Ponticus (ca. 390–after 322 B.C.), so called because he was a native of Heraclea on the Pontus (the Black Sea), was a contemporary of Aristotle's and had also studied at the Academy under Plato. His cosmology differed from that of Plato and Aristotle in at least two fundamental points, possibly because after leaving the Academy he may have studied with the Pythagoreans. The first point of difference concerned the extent of the cosmos, which Heraclides thought to be infinite rather than finite. A second difference concerned the apparent circling of the stars around the celestial pole, which Heraclides said was actually due to the rotation of the earth on its axis in the opposite sense. Simplicius, in his commentary on Aristotle, writes that "Heraclides supposed that earth is in the center and rotates while the heaven is at rest, and he thought by this supposition to save [i.e., account for] the phenomena."

Aristotle was succeeded as head of the Lyceum by his associate Theophrastus (ca. 371–ca. 287 B.C.), to whom he bequeathed his enormous library, which included copies of all his works. Theophrastus is considered to be the second founder of the Lyceum, which he directed for thirty-seven years, reorganizing and enlarging the school.

Theophrastus was as prolific and encyclopedic as Aristotle, and Diogenes Laertius ascribes 227 books to him, most of which are now lost. Two of his extant works, the History of Plants and the Causes of Plants, have earned him the title "Father of Botany," while his book On Stones represents the beginning of geology and mineralogy. His work on human behavior, entitled Characters, is a fascinating description of the types of people living in Athens during his time, all of whom still seem to be represented in the modern city.

Athens went through profound changes during the years that Theophrastus headed the Lyceum. In 322 B.C. the city came under the harsh rule of Antipater, one of the Diadochoi, or Successors, the Macedonian generals who divided up the empire of Alexander after his death. Cassander, another of the Diadochoi, took control of the city in 317 B.C., installing as his governor Dimitrios of Phaleron, who had studied in the Lyceum under Theophrastus. Ten years later Athens was captured by Dimitrios I of Macedonia, son of Antigonus I, another of the Diodochoi. This led to a series of civil wars that lasted for nearly half a century, and in that time the government of Athens changed hands seven times. During that interval Athens began to decline and was eventually surpassed by Alexandria, the new city that had been founded by Alexander in 331 B.C. on the Canopic branch of the Nile.

Aladdin's Lamp

Theophrastus was succeeded as head of the Lyceum by Straton of Lampsacus (d. ca. 268 B.C.), who had been his student. Straton is credited with more than forty works, all of which are lost except for fragments. His most important works were considered to be those on physics, which is what led later writers to call him Straton the Physicist. Diogenes Laertius describes Straton as "a distinguished man who is generally known as 'the physicist,' because more than anyone else he devoted himself to the study of nature."

One of Straton's writings on physics is a lost work entitled *On Motion*, which is discussed in a commentary by Simplicius. According to Simplicius, Straton was the first to demonstrate that falling bodies accelerate—that is, that their velocity increases over time. "For if one observes water pouring down from a roof and falling from a considerable height, the flow at the top is seen to be continuous, but the water at the bottom falls to the ground in discontinuous parts. This would never happen unless the water traversed each successive space more swiftly."

Straton or one of his contemporaries may have written the Aristotelian work entitled *Mechanics*. This contains the earliest extant statement of the law of the lever: if two objects are suspended from a lever, they will balance if their distances from the fulcrum are inversely proportional to their weights.

Two other schools of philosophy were founded in Athens late in the fourth century B.C. These were not formal institutions like the Academy and the Lyceum but more loosely organized groups that met to discuss philosophy. One of the schools, known as the Garden, was founded by Epicurus of Samos (341–270 B.C.) and the other, the Porch, was begun by Zeno of Citium (ca. 335–263 B.C.). The name of the first school stemmed from the fact that Epicurus lectured in the garden of his house, while the second was named for the Stoa Poikile, or Painted Porch, in the agora, the meeting place of Zeno and his disciples, who came to be known as the Stoics. Both Epicurus and Zeno created comprehensive philosophical systems that were divided into three parts—ethics, physics, and logic—in which the last two were subordinate to the first, whose goal was to secure happiness. According to Epicurus: "If we are not troubled with doubts about the heavens, and about the possible meaning of death, and by failures to understand the limits of pain and desire, then we should have no need of natural philosophy."

The physics of Epicurus was based on the atomic theory, to which he added one new concept: that an atom moving through the void could at any instant "swerve" from its path. This eliminated the absolute determinism that had made the original atomic theory of Leucippus and Democritus unacceptable to those who, like the Epicureans, believed in free will. Zeno and his followers rejected the atom and the void, for they looked at nature as a continuum in all of its aspects—space, time, and matter—as well as in the propagation and sequence of physical phenomena. These two opposing schools of thought about the nature of the cosmos—the Epicurean atoms in a void versus the continuum of the Stoics—have competed with each other from antiquity to the present, for they seem to represent antithetical ways of looking at physical reality.

Thus even as Athens was giving way to Alexandria as the intellectual center of the Greek world, it continued to be the School of Hellas, with the creation of two new philosophical systems that would take their place beside those of Plato and Aristotle in their influence on Western thought.

3

HELLENISTIC ALEXANDRIA: THE MUSEUM AND THE LIBRARY

After Alexander's death in 323 B.C., his general Ptolemaeus, better known as Ptolemy, took control of Egypt, ruling at Alexandria with the title of satrap. He declared himself king in 305 B.C., taking the name of Soter, or Savior, beginning a reign of more than twenty years and a dynasty, the House of Ptolemy, that would last for nearly three centuries.

Alexandria soon became a great cultural center under Ptolemy I Soter (r. 305–283 B.C.), who wrote a biography of Alexander the Great. The Alexandrian renaissance he began centered on the establishment of two renowned institutions, the Museum and the Library, which he founded and which were further developed by his son and successor Ptolemy II Philadelphus (r. 283–245 B.C.).

The Museum took its name from the fact that it was dedicated to the Muses, the nine daughters of Zeus and Mnemosyne, goddess of memory, who were the patron deities of the humanities. There were Temples of the Muses elsewhere in the Greek world, including a Museum in the Platonic Academy and one founded by Theophrastus in memory of Aristotle. The Alexandrian Museum and its attached Library were meant to be a university and research center, patterned on the famous schools of philosophy in Athens, most notably the Academy and the Lyceum.

The geographer Strabo, writing in the first quarter of the first century

A.D., notes that the Museum was part of the royal palace complex of the Ptolemies: "The Museum is also a part of the royal palaces; it has a public walk, an exedra with seats, and a large house, in which is the common mess-hall of the men of learning who share the Museum. This group of men not only hold property in common, but also have a priest in charge of the Museum, who formerly was appointed by the kings, but is now appointed by Caesar."

The Museum was more like a research institute than a college, with the emphasis on science rather than the humanities. It would have included an astronomical observatory as well as rooms for anatomical dissection and physiological experiments, and around it were botanical and zoological gardens.

The scientific character of the Museum was probably due to Straton of Lampsacus, the Physicist. Straton moved to Alexandria circa 300 B.C. to serve as tutor to the future Ptolemy II Philadelphus, remaining there until he returned to Athens in 288 B.C. to succeed Theophrastus as director of the Lyceum. The prince developed a deep interest in geography and zoology through his studies with Straton, and this was reflected in his development of the Museum when he succeeded his father as king in 283 B.C.

The organization of the Library was probably due to Dimitrios of Phaleron, the former governor of Athens, who had been forced to flee from the city in 307 B.C., after which he was given refuge in Alexandria by Ptolemy I. Dimitrios, a former student at the Lyceum in Athens, is believed to have been the first chief librarian of the Library, a post he held until 284 B.C. According to Aristeas Judaeus, a Jewish scholar in the reign of Ptolemy II Philadelphus, Dimitrios "had at his disposal a large budget in order to collect, if possible, all the books in the world, and by purchases and transcriptions he, to the best of his ability, carried the king's objective into execution."

This policy continued through the reigns of Ptolemy II Philadelphus and Ptolemy III Euergetes (r. 247–221 B.C.). Athenaeus of Naucratis, who flourished circa A.D. 200, reports that Ptolemy II bought the books of Aristotle and Theophrastus and transferred them to "the beautiful city of Alexandria." By the time of Ptolemy III the Library was reputed to have a collection of more than half a million parchment rolls, including all the great works in Greek science and the humanities from Homer onward. This led the third Ptolemy to build a new branch of the Library within the Serapeum, the temple of Serapis. Epiphanius of Salamis, a Christian of the fourth century A.D., refers to this addition in writing of "the first library and another built in the Serapeum, smaller than the first, which was called the daughter of the first." Most classical authors do not refer to two libraries, but to "the Royal Library," the "great library," or "the libraries," occasionally mentioning "the daughter library" in the Serapeum.

Dimitrios was succeeded as chief librarian by Zenodotus of Ephesus, who held the post until 245 B.C. His principal assistant was the poet Callimachus of Cyrene (ca. 305–ca. 240 B.C.), who classified the 120,000 works of prose and poetry in the library according to author and subject, the first time this had ever been done. His compilation, known as the *Pinakes* (Tables), bore the title *Tables of Persons Eminent in Every Branch of Learning Together with a List of Their Writings,* and filled more than 120 books, five times the length of Homer's *Iliad.*

The only scientist to serve as chief librarian of the Library was Eratosthenes of Cyrene (ca. 275–ca. 195 B.C.), who was appointed by Ptolemy III in or near 235 B.C. and held the post until his death. Renowned as a mathematician, astronomer, and geographer, he also made a study of Old Attic comedy and was the first to write a chronology of Greek history and literature.

Eratosthenes was the first to draw a map of the known world based on a system of meridians of longitude and parallels of latitude, which allowed him to make an accurate measurement of the earth's circumference. This measurement was done by recording simultaneous observations made at Alexandria and Syene, a distance of 5,000 *stades* to the south. It was observed that at the summer solstice the sun was directly overhead at noon in Syene, while on a sundial at Alexandria it cast a shadow equal to one-fiftieth of a circle. Assuming that the sun was so far away that its rays were parallel at Syene and Alexandria, Eratosthenes concluded that the north-south distance between the two places was one-fiftieth of the earth's circumference. Thus the circumference of the earth was fifty times the distance between Syene and Alexandria, or 250,000 *stades.* The precise value used by Eratosthenes for the length of a *stade* is unknown, so it is not possible to evaluate the accuracy of his result, but it was certainly of the right order of magnitude.

The great school of mathematics at Alexandria was founded by

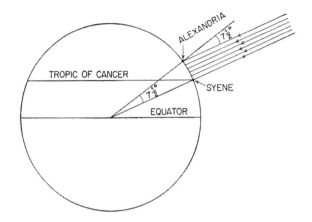

Measurement of the earth's circumference by Eratosthenes.

Euclid (fl. ca. 295 B.C.), who is said to have taught there early in the third century B.C., but very little else is known of his life. The philosopher Proclus, writing in the fifth century A.D., says that Euclid "lived in the time of the first Ptolemy" and that he was "younger than the students of Plato but older than Eratosthenes and Archimedes."

Euclid is renowned for his *Elements of Geometry*, the earliest extant textbook on the subject, still in use today. A modern historian of science notes that the *Elements* "has exercised an influence upon the human mind greater than that of any other work except the Bible."

The *Elements* established the foundations not only of plane geometry, but also of algebra and number theory. Proclus says that Euclid composed the *Elements* by "collecting many of the theorems of Eudoxus, perfecting many of those of Theaetetus [a pupil of Plato's], and also bringing to irrefutable demonstration the things which are somewhat loosely proved by his predecessors."

Aside from the mathematical content of the *Elements*, one of its most important qualities is the logical form and order in which the theorems are presented, for this was to be the model for all future work in Greek mathematics and mathematical physics. Equally important is the axiomatic nature of the *Elements*, all of geometry following as a logical deduction from a few assumptions, themselves taken as necessarily true, which when applied to physics and astronomy represented the Platonic geometrization of nature.

Euclid's extant writings include several other works on mathematics as well as a textbook on astronomy, the *Phaenomena*, and an elementary treatise on perspective, the *Optica*. The *Optica* was the first Greek work on the subject and the only one until Claudius Ptolemaeus wrote his treatise on optics in the mid-second century A.D. One of the assumptions made by Euclid in the *Optica* is that vision involves light rays proceeding in straight lines from the eye to the object. This erroneous view, known as the extramission theory, was accepted by most writers on optics until the seventeenth century.

Greek mathematical physics reached its peak with the works of Archimedes (ca. 287–212 B.C.), who was born at Syracuse in Sicily. Archimedes is said to have spent some time in Egypt, and he corresponded with Eratosthenes. It is probable that he studied in Alexandria under the successors of Euclid, for he was certainly familiar with the *Elements* and quoted from it extensively.

Archimedes was a relative and friend of Hieron II of Syracuse and of the king's son and successor Gelon II. He worked for both Hieron and Gelon as a military engineer, constructing devices such as catapults, burning mirrors, and a system of compound pulleys for moving large ships with minimum effort. These devices were used to great advantage by the Syracusans in resisting the siege of their city in 212 B.C. by the Roman general Marcellus. But Marcellus eventually captured the city and Archimedes was killed by a Roman soldier, supposedly while he was drawing a geometrical proposition in the sand.

Archimedes was famous for his inventions, one of which was an orrery, a working model of the celestial motions. Cicero actually saw this orrery, which he claimed showed the motion of the sun and moon and demonstrated both solar and lunar eclipses. Another of his inventions, known as Archimedes' screw, is still used in Egypt to lift water in primitive irrigation systems. Plutarch remarks that Archimedes himself did not have high regard for his inventions, regarding them as being merely the "diversions of a geometry at play."

Plutarch writes of Archimedes' mathematical demonstrations that "it is not possible to find in geometry more difficult and troublesome questions or proofs set out in simpler and purer terms." One can appreciate the nature of Archimedes' works from the titles of some of his treatises, such as *On the Measurement of a Circle, On the Sphere and the Cylinder, On the Equilibrium of Planes, On Floating Bodies*, and *The Sand Reckoner*.

In the first of these treatises Archimedes rigorously determined the surface area of a circle using the so-called method of exhaustion, whereby he obtained successively better approximations by computing the areas of regular polygons inside and outside the circle. He used this method in his other mathematical works to measure the areas and volumes of various figures, such as in his treatise *On the Sphere and the Cylinder*. There, considering a cylinder circumscribing a sphere, he found that the ratio of their areas was 3:2, and he was so proud of this discovery that he had the figure carved on his tomb.

The treatise *On the Equilibrium of Planes* deals with statics, the study of mechanical systems in equilibrium. There Archimedes uses the law of the lever to find the so-called center of gravity of various figures—that is, the point at which all of their weight is in effect concentrated. The problems are idealized, neglecting friction and other extraneous factors, and the treatment is completely deductive and geometrical, patterned on Euclid's *Elements*. Archimedes' work on the lever gave rise to his legendary boast to King Hieron: "Give me a place to stand and I shall move the earth."

The treatise *On Floating Bodies* applies the same kind of geometric analysis to hydrostatics, the study of fluids in equilibrium. The basic proposition that he uses here is the famous Archimedes' principle, which says that a body wholly or partially immersed in a fluid is buoyed up by a force equal to the weight of fluid displaced. The first-century Roman writer Vitruvius tells the story that Archimedes discovered this principle when he entered a bathtub and noticed the increasing sense of buoyancy as he immersed himself and the water level rose. According to Vitruvius, Archimedes "without a moment's delay, and transported with joy . . . jumped out of the tub and rushed home naked, crying in a loud voice that he had found what he was seeking; for as he ran he shouted repeatedly in Greek, 'Heureka!' [I have found it]."

Vitruvius goes on to tell how Archimedes used his principle to solve a practical problem: determining whether a golden crown made for King Hieron had been adulterated with another metal. He weighed the crown in water and found that it displaced a greater volume of water than the same weight of pure gold. This showed that the crown was less dense than pure gold, and thus that it had been made with an admixture of a lighter metal. Archimedes had discovered the concept of specific gravity, the weight of a body relative to that of an equivalent volume of water.

The Sand Reckoner is dedicated to King Gelon, to whom Archimedes explains a method that he had developed for expressing extremely large numbers. This was virtually impossible with the system then used by the Greeks, where numbers were written in terms of the letters of the alphabet. As an example, Archimedes gives the number of grains of sand in "a volume equal to that of the cosmos," that is, "the sphere whose center is the center of the earth and whose radius is the distance between the center of the sun and the center of the earth." He then makes reference to a new theory that had been proposed by Aristarchus of Samos, an older contemporary.

> Aristarchus of Samos has, however, enunciated certain hypotheses in which it results from the premises that the universe is much greater than that just mentioned. As a matter of fact, he supposes that the fixed stars and the sun do not move, but that the earth revolves in the circumference of a circle about the sun, which lies in the middle of the orbit, and that the sphere of the fixed stars, situated about the same center as the sun, is so great that the circle in which the earth is supposed to revolve has the same ratio to the distance of the fixed stars as the center of the sphere to its surface.

The last sentence is of particular significance, for it explains why there is no stellar parallax, or apparent displacement of the stars, when the earth moves in orbit around the sun in the heliocentric theory of Aristarchus. It posits that even the nearest stars are so far away, compared to the radius of the earth's orbit around the sun, that their parallax is far too small to be detected by the naked eye. In fact, this effect was not observed until the mid-nineteenth century, by which time telescopes of sufficient resolving power had been developed.

Aristarchus of Samos (ca. 310–ca. 230 B.C.) was a student of Straton the Physicist's, probably at the Lyceum in Athens. The only work of Aristarchus's that has survived is his treatise *On the Sizes and Distances of the Sun and the Moon*. Here the radii of the sun and moon relative to that of the earth and their distances in earth radii were calculated from geo-

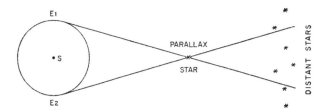

Stellar parallax. E1 and E2 represent two positions of the earth six months apart in its orbit around the sun S. The size of the earth's orbit is greatly exaggerated. The parallax star is one that is much closer than the distant stars, so that it will be displaced with respect to them as observed from the earth around the two positions shown. The farther away the nearby star is, the smaller will be its angle of parallax.

metrical considerations. The first observation was that the sun and the moon appear to be the same size, indicating that their diameters must be proportional to their distances from the earth. The second was a measurement of the lunar dichotomy (i.e., the angular separation of the sun and moon at half-moon), and the third was an estimation of the breadth of the earth's shadow where the moon passes through it at the time of a lunar eclipse. The results of these measurements led Aristarchus to conclude that the sun is about 19 times farther from the earth than the moon, and that the sun is approximately 6¾ times as large and the moon about ⅓ as large as the earth. All of his values are grossly underestimated, because of the crudeness of his observations, but his geometrical methods were sound.

The only ancient astronomer known to have accepted the heliocentric theory of Aristarchus was Seleucus the Babylonian, who lived in the second century B.C. One of the reasons for the lack of acceptance was that the theory conflicted with general religious belief, which had the earth as the stationary center of the cosmos. Cleanthes of Assos, who flourished in the mid-third century B.C., wrote a tract condemning the theory, in which he said that Aristarchus not only had the earth in orbit around the sun but also had it rotating on its axis. In this work, which is quoted by Plutarch, Cleanthes remarks that Aristarchus should be charged with impiety "on the ground that he was disturbing the hearth of the universe because he sought to save [the] phenomena by supposing that the heaven is at rest while the earth is revolving along the ecliptic and at the same time is rotating about its own axis."

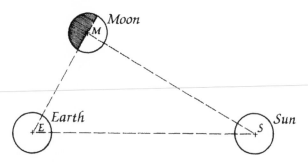

From Aristarchus's *On the Sizes and Distances of the Sun and the Moon*. *Above:* The lunar dichotomy. *Below:* A lunar eclipse diagram.

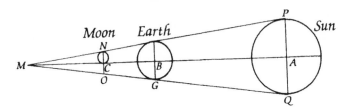

The only Hellenistic mathematician comparable to Archimedes is his younger contemporary Apollonius of Perge. Apollonius was born circa 262 B.C. in Perge, on the Mediterranean coast of Asia Minor, and in his youth he was sent to study in Alexandria, where he flourished during the reigns of Ptolemy III and Ptolemy IV Philopator (r. 221–203 B.C.). He was also an honored guest in the court of Attalos I (r. 241–197 B.C.) of Pergamum in northwestern Asia Minor, which had become a center of Greek culture, renowned for its library.

The only major work of Apollonius's that has survived is his treatise *On Conics*, though even there the last book is lost. This was the first comprehensive and systematic analysis of the three types of conic sections: the ellipse (of which the circle is a special case), the parabola, and the hyperbola.

Apollonius is also credited with formulating mathematical theories to explain the apparent retrograde motion of the planets. One of the theories has a planet moving around the circumference of a circle, known as the epicycle, whose center itself moves around the circumference of another circle, called the deferent, centered at the earth. The second theory has a planet moving around the circumference of an eccentric circle whose center does not coincide with the earth. He also

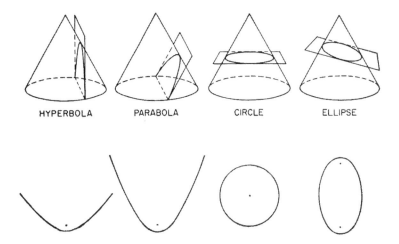

HYPERBOLA   PARABOLA   CIRCLE   ELLIPSE

The conic sections.

showed that the epicycle and eccentric circle theories are equivalent, so that either model can be used to describe retrograde planetary motion. Aside from the great theoreticians of the Hellenistic era, there were also a number of gifted inventors whose works were extremely influential in the development of applied science.

Ctesibius of Alexandria, the son of a barber, who is believed to have flourished ca. 270 B.C., was renowned as an inventor of toys and devices involving pneumatics—air under pressure. He is credited with inventing a force pump, a war catapult, a fire engine, a water clock, a hydraulic organ, and a singing statue, which he made for the empress Arsinoë, the sister and wife of Ptolemy II. The most elaborate of his water clocks told the hours with a succession of figures known as *parerga*, such as moving puppets and whistling birds, the ancestor of the cuckoo clock. All of his written works are now lost, but his ideas and inventions were revived by his two most notable followers, Philo of Byzantium and Hero of Alexandria.

Philo of Byzantium flourished circa 250 B.C. His extant writings comprise three books from a large work on mechanics: *On Catapults, On Pneumatics,* and *On Besieging and Defending Towns.* In the first of these books Philo states that he traveled to Alexandria, where people described to him the bronze-spring catapult built by Ctesibius. The second book describes a number of demonstrations almost certainly taken

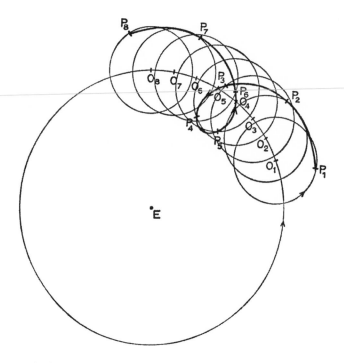

The epicycle theory of Apollonius; the model was used to explain the retrograde motion of a planet P.

from Ctesibius, including several pneumatic toys. The third of the three books deals with the besieging of walled towns, including the use of catapults and other engines of war, as well as stratagems such as secret messages, cryptography, and poisons.

Hero of Alexandria flourished circa A.D. 62. The longest of his extant works by far is his *Pneumatica*, described by one modern historian as containing "almost exclusively apparatuses for parlor magic." One is his famous steam engine, in which a glass bulb is made to rotate by two jets of steam escaping from it in opposite directions at either end of a diameter. The first chapters, which derive largely from Philo, describe experiments to show that air is a body and that it is possible to produce a vacuum, contrary to the Aristotelian doctrine. One of the demonstrations described by Hero may derive directly from Straton the Physicist. "Thus," he writes, "if a light vessel with a narrow mouth be taken and applied to the lips, and if the air be sucked out and discharged, the vessel will be suspended from the lips, the vacuum drawing the flesh towards it so that the evacuated vessel be filled. It is manifest from this that there was a continuous vacuum within the vessel."

Other inventions are described in Hero's treatise *On Automata-Making*, most notably the *thaumata*, or "miracle-working" devices such as one that opened and closed the doors of a temple using air pressure. The most elaborate of the automata are two puppet shows, one of them showing Dionysus pouring out a libation in front of a temple while bacchants dance about him to the sound of trumpets and drums, another representing a naval battle in which Athena destroys the ships of Ajax with thunder and lightning. Hero also made important contributions in optics as well as applied mathematics.

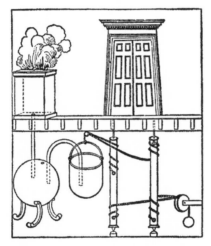

Inventions of Hero's: *Above:* Temple doors opened by fire on an altar. *Below:* A steam engine.

Hipparchus, the greatest observational astronomer of antiquity, was born in Nicaea, in northwestern Asia Minor. His life span can be estimated from the dates of his earliest known observation, the autumnal equinox of 26–27 September 147 B.C., and the latest, a lunar position on 7 July 127 B.C. It is probable that he spent the latter part of his career in Rhodes, where he is known to have made observations from 141 to 127 B.C. What little is known of his life comes from the geographer Strabo, who says that Hipparchus made use of the Library at Alexandria, and from the astronomer Ptolemy, who refers frequently to his observations and often quotes him directly.

All of the writings of Hipparchus have been lost except for his first work, a commentary on the *Phainomena* of Aratus of Soli (ca. 310–ca. 240 B.C.), a Greek poem describing the constellations. This commentary served to popularize the names of stars and constellations, many of which have been perpetuated in the modern world. It contained a catalog of some 850 stars, for each of which Hipparchus gave the celestial coordinates, including those of a "nova," or new star, which suddenly appeared in 134 B.C. in the constellation Scorpio. He also estimated the brightness of the stars, assigning to each of them a "magnitude," which equaled 1 for the brightest stars and 6 for the faintest, a system still used in modern astronomy.

One of the lost writings of Hipparchus is a book on the sizes and distances of the sun and the moon, in which he apparently made a great improvement over Aristarchus. These and other measurements and theories of Hipparchus's were used by Ptolemy, who paid due credit to his predecessor.

Hipparchus is also renowned for his discovery of the precession of the equinoxes—that is, the slow movement of the celestial pole in a circle about the perpendicular to the ecliptic. The earth's precession manifests itself as a gradual advance of the spring equinox along the ecliptic, thus causing a progressive change in the celestial longitude of the stars. Hipparchus discovered this effect by comparing his star catalog with observations made 128 years earlier by the astronomer Timocharis, which led him to conclude that the celestial longitude of the star Spica in the constellation Virgo had changed by 2 degrees in that interval of time, amounting to an annual precession of 45.2 seconds of arc. This allowed him to make an accurate determination of the length of the so-called

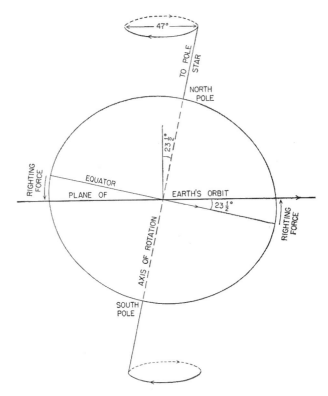

The precession of the equinoxes. *Above:* Precession caused by the forces of the sun and the moon on the earth's equatorial bulge. *Below:* The path of the north celestial pole in the celestial sphere. The celestial pole describes a radius of 23.5 degrees around the ecliptic pole.

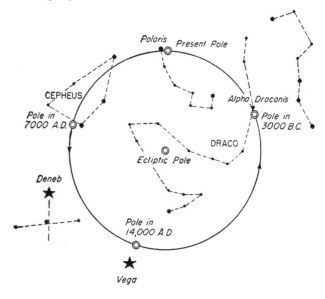

tropical year by measuring the time between two summer solstices, one observed by Aristarchus in 280 B.C. and another by himself in 135 B.C. The value that he found was equivalent to 365.2467 days, a significant improvement over the old value of 365.25 days, which did not take into account the precession of the equinoxes. The currently accepted value of the tropical year is 365.242190 days, which means that the measurement by Hipparchus was in error by somewhat less than 1 part in 10,000.

Hipparchus is also celebrated as a mathematician, his great achievement being the development of spherical trigonometry, which he applied to problems in astronomy.

Theodosius of Bithynia, a younger contemporary of Hipparchus's, is remembered for his *Sphaerica,* a treatise on the application of spherical geometry to astronomy, which was translated into Arabic and later into Latin, remaining in use until the seventeenth century.

When Ptolemy XII died, in 51 B.C., his eldest daughter, Cleopatra VII, succeeded to the throne, while her younger brother Ptolemy XIII became coruler. A civil war began between Cleopatra and her brother, both of whom were captured by Caesar in 48 B.C. after his victory over Pompey, who was assassinated in Egypt. Caesar's small army was attacked by a much larger Egyptian force; during the fighting a fire broke out and destroyed many buildings in the harbor quarter of Alexandria, including at least part of the Library. Seven years later Mark Antony promised Cleopatra that he would make up the loss by giving her some 200,000 volumes from the library of Pergamum. In any event, it appears that the Library of Alexandria survived the fire, for there are several mentions of it in the imperial Roman era.

The Ptolemaic dynasty came to an end in 30 B.C. when Cleopatra committed suicide in Alexandria after she and Antony had been defeated by Octavian at the Battle of Actium. Alexandria then came under the rule of Rome, whose imperial era began when Octavian became Augustus in 27 B.C.

The lifetime of the Greek geographer Strabo (63 B.C.–ca. A.D. 25) extended from the end of the Ptolemaic period through the first half century of the Roman imperial era. He was born at Amaseia in the Pontus, and in his youth he studied first at Nysa in Asia Minor, then in Alexandria, and later in Rome. He lived for a long time in Alexandria, where he would likely have studied the works of Eratosthenes and other Greek geographers, whom he mentions. His earlier historical work is lost, but his more important seventeen-book *Geography* has survived. Strabo followed the tradition of Eratosthenes in geography, but added encyclopedic descriptions "of things on land and sea, animals, plants, fruits, and everything else to be seen in various regions." He said that the northern limit of the inhabited world was Ireland, which he called Ierne, "the home of men who are complete savages and lead a miserable existence because of the cold." He went on to say that "its inhabitants are more savage than the Britons, since they are man-eaters as well as herb-eaters, and since, further, they count it an honorable thing, when their fathers die, to devour them, and openly to have intercourse, not only with the other women, but with their mothers and sisters."

Another who came to study in Alexandria was Dioscorides Pedanius (fl. A.D. 50–70), from Anazarbus in southeastern Asia Minor, who later became a physician in the Roman army during the reigns of Claudius (r. A.D. 41–54) and Nero (r. A.D. 54–68). Dioscorides is regarded as the founder of pharmacology, renowned for his *De Materia Medica*, a systematic description of some six hundred medicinal plants and nearly a thousand drugs.

Nicomachus of Gerasa (fl. ca. A.D. 100) is noted for his *Introduction to Arithmetic*. This was an elementary handbook on those parts of mathematics that were needed for an understanding of Pythagorean and Platonic philosophy. The book has many errors and other shortcomings, but it was influential until the sixteenth century, giving Nicomachus the undeserved reputation of being a great mathematician.

Ancient Greek astronomy culminated with the work of Claudius Ptolemaeus (ca. A.D. 100–ca. 170), known more simply as Ptolemy. All that is known of Ptolemy's life is that he flourished in Alexandria during the successive reigns of the emperors Hadrian (r. A.D. 117–38) and Antoninus Pius (r. A.D. 138–61). The most influential of his writings is his *Mathematical Synthesis*, better known by its Arabic name, the *Almagest*, the most comprehensive work on astronomy that has survived from antiquity.

The topics in the *Almagest* are treated in logical order through thirteen books. Book I begins with a general discussion of astronomy, including Ptolemy's view that the earth is stationary "in the middle of the heavens." The rest of Book I and all of Book II are devoted principally to the development of the spherical trigonometry necessary for the whole work. Book III deals with the motion of the sun and Book IV with lunar motion, which is continued at a more advanced level in Book V, along with solar and lunar parallax. Book VI is about eclipses; Books VII and VIII are about the fixed stars; and Books IX through XIII are on the planets.

Ptolemy's trigonometry and catalog of stars are based on the work of Hipparchus, and his theory of epicycles and eccentrics is derived from Apollonius. The principal modification made by Ptolemy is that the center of the epicycle moves uniformly with respect to a point called the equant, which is displaced from the center of the deferent circle, a device that was to be the subject of controversy in later times.

The extant writings of Ptolemy also include other treatises on astronomy, the *Handy Tables, Planetary Hypotheses, Phases of the Fixed Stars, Analemma,* and *Planisphaerium;* a work on astrology called the *Tetrabiblos;* and treatises entitled *Optics, Geography,* and *Harmonica,* the latter devoted to musical theory.

Ptolemy's *Tetrabiblos* is the classic Greek work on astrology, the pseudoscience of astronomical divination based on the notion that celestial bodies influence human affairs. Ptolemy was himself skeptical about some of the credulous beliefs involved in astrology, as when he states in

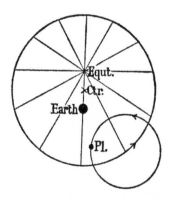

Ptolemy's concept of the equant. In the diagram the planet (Pl.) moves on an epicycle whose center describes an eccentric circle, i.e., one whose center (Ctr.) is outside the earth. The center of the epicycle moves uniformly with respect to the equant (Equt.). The equant and the earth are equidistant from the center of the eccentric circle on opposite sides of a diameter.

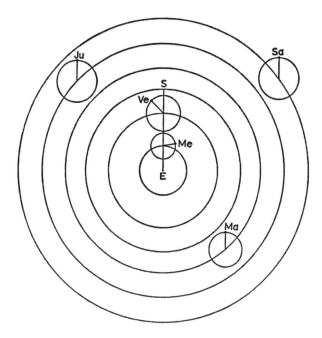

A simplified version of Ptolemy's planetary model (ignoring eccentricity and equants and not to scale) to show the relationship between the planets and the sun in Ptolemy's system. The centers of the epicycles of Mercury (*Me*) and Venus (*Ve*) are on the line joining the earth *E* to the sun *S*. For the outer planets—Mars (*Ma*), Jupiter (*Ju*) and Saturn (*Sa*)—the line joining the planet to the center of its epicycle remains parallel to *ES*.

Book III of the *Tetrabiblos* that "we shall dismiss the superfluous non-sense of the many, that lacks any plausibility, in favor of the primary natural causes."

Ptolemy's researches on light are presented in his *Optics*. Here he gives the correct form for the law of reflection, already known to Euclid, that the incident and reflected rays make the same angle with the surface normal, the perpendicular to the mirror at the point of incidence. His experiments led him to an empirical relation for the law of refraction, the bending of light when it passes from one medium to another. He found that when light passes into a denser medium, as from air to water or glass, the refracted ray makes a smaller angle with the surface normal than does the incident ray. He then used these laws to find the location, size, and form of images produced by reflection and refraction.

Ptolemy's *Geography* is the most comprehensive work on that subject to survive from the ancient world. One defect of his work is that his value for the circumference of the earth is too small by a factor of one-third. The most conspicuous error in his map of the *ecumenos*, or the inhabited world, is the extension of the Eurasian landmass over 180 degrees of longitude instead of 120 degrees. Nevertheless, Ptolemy's treatise was by far the best geographical work produced in antiquity.

Galen of Pergamum (A.D. 130–after 204), the most renowned medical writer of antiquity, was a younger contemporary of Ptolemy's. Galen was born in Pergamum and studied there as well as in Smyrna, Corinth, and Alexandria. He served his medical apprenticeship at the healing shrine of Asclepios at Pergamum, where his work treating wounded gladiators gave him an unrivaled knowledge of human anatomy, physiology, and neurology. In 161 he moved to Rome, where he spent most of the rest of his life, serving as physician to the emperors Marcus Aurelius (r. A.D. 161–80), Lucius Verus (coemperor; r. A.D. 161–69), and Commodus (r. A.D. 180–92).

Galen's writings, translated successively into Arabic and Latin, served as the basis for explaining the anatomy and physiology of the human body until the seventeenth century, earning him the title of "Prince of Physicians." The title of one of his treatises is *That the Best Doctor Is Also a Philosopher*. His philosophical bent is evident in his medical writings,

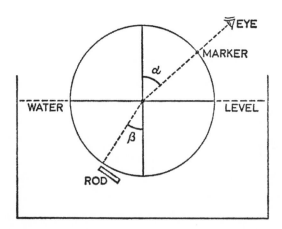

Ptolemy's experimental investigation of refraction, where α is the angle of incidence and β the angle of refraction.

where he interprets the work of Plato, Aristotle, Epicurus, and others, and also in his treatises *On Scientific Proof* and *Introduction to Logic*. He wrote on psychology as well, including an imaginative analysis of dreams, seventeen centuries before Freud. One of the psychic complaints recognized by Galen was lovesickness, which he believed to be the principal cause of insomnia, and he notes that "the quickening of the pulse at the name of the beloved gives the clue."

The last great mathematician of antiquity was Diophantus of Alexandria (fl. ca. A.D. 250), who did for algebra and number theory what Euclid had done for geometry. Little is known of his life beyond the bare facts given in an algebraic riddle about Diophantus in the *Greek Anthology*, dating from the fifth or sixth century: "God granted him to be a boy for the sixth part of his life, and adding a twelfth part to this He clothed his cheeks with down; He lit him the light of wedlock after a seventh part, and five years after his marriage He granted him a son. Alas! Late-born wretched child; after attaining the measure of half his father's life, chill Fate took him. After consoling his grief by this science of numbers for four years he ended his life."

Assuming that the biographical facts in this riddle are correct, Diophantus lived to be eighty-four years old. His most important treatise is the *Arithmetica*, of which six of the original thirteen books have survived. This work was translated into Latin in 1621; six years later it inspired the French mathematician Pierre Fermat to create the modern theory of numbers. After reading Diophantus's solution of the problem of dividing a square into the sum of two squares, which is the Pythagorean theorem, Fermat made a notation in the margin of his copy of the *Arithmetica*, stating, "It is impossible to divide a cube into two other cubes, or a fourth power or, in general, any number which is a power greater than the second into two powers of the same denomination." He then noted that he had discovered a remarkable proof "which this margin is too narrow to contain." He never supplied the proof of what came to be called "Fermat's last theorem," which was finally solved in 1995 by Andrew Wiles, a British mathematician working at Princeton University, the last link in a long chain of mathematical development that began in ancient Alexandria. The work of Diophantus is still a part of modern mathematics, studied under the heading of "Diophantine analysis."

Pappus of Alexandria, who flourished in the first half of the fourth century A.D., is renowned for his work in mathematics, astronomy, music, and geography. His treatise entitled *Synagogue* (Collection) is the principal source of knowledge of the accomplishments of many of his predecessors in the Hellenistic era, most notably Euclid, Archimedes, Apollonius, and Ptolemy. His work in mathematics influenced both Descartes and Newton, and one of his discoveries, known today as the theorem of Pappus, is still taught in elementary calculus courses. In one of his works in mathematics he reflects on mathematics in nature, remarking that "Bees . . . by virtue of a certain geometrical forethought . . . know that the hexagon is greater than the square and the triangle and will hold more honey for the same expenditure of material."

An oath signed by "Pappus, philosopher" in a collection of works on alchemy has been attributed to Pappus of Alexandria. If so, it may shed light on his religious views, which would seem to be a mixture of Christian, Pythagorean, alchemical, and astrological beliefs. It reads: "I therefore swear to thee, whoever thou art, the great oath, I declare God to be one in form, but not in number, the maker of heaven and earth, as well as the tetrad of the elements and things formed from them, who has furthermore harmonized our rational and intellectual souls with our bodies, who is borne upon the chariots of the cherubim and hymned by angelic throngs."

During the Hellenistic period the pseudosciences of alchemy and astrology developed strong connections with magic, a mystical influence that would be passed on to Islam and medieval Europe. Much of the resulting lore is found in the so-called Corpus Hermeticum, a collection of writings on alchemy, astrology, and magic that takes its name from the legendary Hermes Trismegestus (Thrice Greatest), a syncretization of the Greek god Hermes and the Egyptian divinity Thoth. Most of these writings, once thought to be of ancient Egyptian origin, are now dated to about the second century A.D. The earliest full description of the Corpus Hermeticum is in the *Stromata* of Clement of Alexandria (ca. 150–ca. 220), who says that of the forty-two books four are astrological and the remainder religious, philosophical, and medical, all of them with a touch of the occult, as in one fragment that states that "philosophy and magic nourish the soul."

The library in the Serapeum survived almost to the end of the fourth century, by which time the Museum seems to have vanished. The emperor Theodosius I issued a decree in 391 calling for the destruction of all pagan temples throughout the empire. Theophilus, bishop of Alexandria, took this opportunity to lead his fanatical followers in demolishing the Serapeum, including its library.

The last contemporary description of the Library is by Aelius Festus Aphthonius, who sometime after 391 paid tribute to its role in making Alexandria the center of Greek culture: "On the inner side of the colonnade were built rooms, some of which served as book stores and were open to those who devoted their life to the cause of learning. It was these study rooms that exalted the city to be the first in philosophy. Some other rooms were set up for the worship of the old gods."

The last scholar known to have worked in the Museum and Library was Theon of Alexandria, who in the second half of the fourth century wrote commentaries on Euclid's *Elements* and *Optica*, as well as on Ptolemy's *Almagest* and *Handy Tables*. A passage in Theon's commentary on the *Handy Tables* led to an interesting development in Arabic astronomy. This is where Theon states that "certain ancient astrologers" believed that the points of the spring and fall equinox oscillate back and forth along the ecliptic, moving through an angle of 8 degrees over a period of 640 years. This erroneous notion was revived in the "trepidation theory" of Arab astronomers, and it survived in various forms into the sixteenth century, when it was discussed by Copernicus.

Theon was the father of Hypatia (ca. 370–415), the first woman to appear in the history of science. Hypatia was a professor of philosophy and mathematics, and circa 400 she became head of the Platonic school in Alexandria. She revised the third book of Theon's commentary on Ptolemy's *Almagest,* and she also wrote commentaries on the works of Apollonius and Diophantus, now lost. Her lectures on pagan philosophy aroused the anger of Saint Cyril, bishop of Alexandria, who in 415 instigated a riot by fanatical Christians in which Hypatia was killed.

The last pagan to head the Platonic school in Alexandria was Ammonius, who directed it from 485 until his death circa 517–26. A distinguished philosopher, astronomer, and mathematician, he was noted for his commentaries on Aristotle. He managed to remain on good terms with the Christian authorities in Alexandria, though he and several of his faculty were pagans. One of his two most famous students, the philosopher John Philoponus, who succeeded him as head of the Platonic school, was a Christian, probably from birth. His other famous student, Simplicius, renowned for his commentaries on Aristotle, seems to have remained a pagan.

Eutocius of Ascalon (born ca. 480) was also a student of Ammonius's, to whom he dedicated his commentary on the first book of Archimedes' *On the Sphere and the Cylinder*. Eutocius later wrote commentaries on two more works of Archimedes'—*On the Measurement of a Circle* and *On the Equilibrium of Planes*—as well as on the first four books of the *Conics* of Apollonius. His commentaries proved to be crucial in the survival of these works.

Ammonius was the last pagan philosopher of Alexandria, for by his time Christianity had triumphed over the old gods who had been worshipped in the Serapeum and the Greek philosophers whose works had been studied in the Museum and the Library. As Tertullian had written two centuries earlier in attacking pagan philosophy, rejecting research in favor of revelation:

> What then has Athens to do with Jerusalem, the Academy with the Church, the heretic with the Christian? Our instruction comes from the Porch of Solomon who himself taught us that the Lord is to be sought in the simplicity of one's heart. . . . We have no need of curiosity after Jesus Christ, nor of research after the gospel. When we believe, we desire to believe no more. For we believe this first, that there is nothing else that we believe.

The ruins of the ancient Library of Alexandria were unearthed by Italian archaeologists in the 1990s. By that time a project had already begun to re-create the ancient Library near its original site, sponsored by UNESCO. This led to the creation of a new library known as the Bibliotheca Alexandrina, which opened on 16 August 2002. The collection of the new institution includes all of the extant works of those who, like Euclid and Archimedes, were associated with the Ptolemaic Library, some of them in the medieval Arabic and Latin translations in which they were transmitted to western Europe, strands in the Ariadne's thread that links the scientific thought of the ancient and modern worlds.

FROM ATHENS TO ROME,
CONSTANTINOPLE, AND
JUNDISHAPUR

P hilosophy and science never flourished in Rome to the extent they did in Athens and Alexandria. Nevertheless, philosophers and scientists of the Roman era included some who were quite influential in the intellectual development of medieval Europe.

By the mid-second century B.C. Rome controlled most of the Greek-speaking world, and Roman armies conquered the remainder in the century that followed. At the same time Rome had been absorbing Greek culture, as evidenced by the development of Roman literature based on Hellenic models.

This cultural exchange accelerated after 155 B.C., when an Athenian embassy arrived in Rome to appeal an unfavorable decision made by Greek arbitrators in a dispute with the city-state of Oropus. The ambassadors were the directors of three of the four renowned philosophical schools of Athens: Carneades of Cyrene, head of the Academy; Critolaos of Phaselis, director of the Lyceum; and Diogenes of Babylon, leader of the Stoic school. The appeal was unsuccessful and the ambassadors returned to Athens, having achieved nothing other than the stimulation of philosophical discourse in Rome. The Epicurean school was not represented in the embassy, probably because in their pursuit of happiness its members chose to avoid involvement in public life.

A decade later the Stoic philosopher Panaetius of Rhodes (ca. 185–109 B.C.) came from Athens to Rome, where he mostly remained until he returned to Athens in 129 B.C. to head the Stoa. Cicero, who was deeply influenced by Panaetius, writes of him that "he fled from the gloom and harshness [of the rigorous Stoics] and did not approve of their thorny arguments. In one branch of philosophy [ethics] he was more gentle, in the other [physics and logic] clearer. He was always quoting Plato, Aristotle, Xenocrates, Theophrastus, Dicaearchus, as his writings clearly show."

Posidonius, one of last original thinkers of the Stoic school, studied at Athens under Panaetius before moving to Rhodes. There he attracted students from Rome, including Cicero, and was visited by leading Romans during their eastern journeys, most notably Pompey. He represented the Rhodians on an embassy to Rome in 87 B.C., and his extensive travels took him as far as Gadeira (Cadiz) in Spain, where he observed the tides of the Atlantic. He described the tides in his treatise *On Ocean*, where he ascribed the phenomenon to the combined actions of sun and moon. None of his wide-ranging works survive other than in fragments, though he was extremely influential in the subsequent intellectual history of the Roman world and of medieval Europe.

The Roman poet Lucretius (ca. 94–ca. 50 B.C.) is said to have followed the Epicurean maxim "Live unnoticed" so carefully that virtually nothing is known about his life. St. Jerome reports that Lucretius was born in 94 B.C.; that an overdose of an aphrodisiac drove him mad; that in his intervals of sanity he wrote several books, which were later edited by Cicero; and that he committed suicide at the age of forty-four.

Lucretius is renowned for a superb didactic poem in six books entitled *De Rerum Natura* (On the Nature of Things), based on the atomic theory and teaching of Epicurus. In the first book Lucretius reveals his intention of using the atomic theory to overthrow "this very superstition which is the mother of sinful and impious deeds." He denies creation on the grounds of the permanence of atomic matter, stating, "Nothing is ever produced by divine power out of nothing." The void is a frame of reference for atoms in motion and "time by itself does not exist; but from things themselves there results a sense of what has already taken place, what is now going on and what is to ensue." We must rely on our senses in studying nature, for, as he asks, "What can be a surer guide as to the distinction of true from false than our senses?"

The second book of *De Rerum Natura* deals with the kinetics of the atomic theory, with the random movements and collisions of the atoms bringing about the groupings and separations that form the various bodies found in nature. Lucretius followed Epicurus in saying that the reason the atoms collide is that "at quite indeterminate times and places they swerve ever so little from their course, just so much that you can call this a change in direction." This was later interpreted to mean that the atomic theory was not deterministic, since the "swerve" made atomic motion unpredictable. The atomic theory was in this way made more acceptable for Christians, since it allowed for free will in human actions. Thus *De Rerum Natura* became popular in medieval Europe, which eventually led to the revival of the atomic theory in the seventeenth century.

The Roman architect and engineer Vitruvius Pollio flourished in the first century B.C. Virtually nothing is known of his life other than the internal evidence in his only known work, *De Architectura*, a treatise on architecture and engineering based partly on his own studies and partly on those of earlier architects, mostly Greeks. The treatise is encyclopedic in scope, dealing not only with the history and principles of architecture, but also with military and civil engineering as well as physics and astronomy, including the construction of sundials. The work of Vitruvius was revived in the European Renaissance and has been influential in architectural studies from then until the present day.

Pliny the Elder (ca. 23–79) was born at Comum (Como) and was probably educated in Rome. He was in command of the Roman fleet sent to evacuate refugees during the eruption of Mount Vesuvius in A.D. 79, when he was asphyxiated by the volcanic fumes. His only surviving work is his *Natural History;* he notes in the preface "that by perusing about 2,000 volumes we have collected in 36 volumes about 20,000 noteworthy facts from one hundred authors we have explored." Pliny the Younger, his nephew and adopted son, described the *Natural History* as "a diffuse and learned work, no less rich in variety than nature itself." The *Natural History* was widely known in medieval Europe, where, despite its uneven quality and generally low level, it represented a large fraction of the scientific knowledge available at that time.

In the opening chapters of Book XXX of the *Natural History*, Pliny gives an account of the origin and history of magic, which he says was disseminated in the Greek world through the works of Pythagoras,

Empedocles, Democritus, and Plato. He goes on to note of magic: "No one should wonder that its authority has been very great, since alone of the arts it has embraced and united with itself the three other subjects which make the greatest appeal to the human mind"—medicine, religion, and divination, particularly astrology—"since there is no one who is not eager to learn the future about himself and who does not think that this is most truly revealed by the sky."

The Roman writer Seneca, a Stoic who flourished in the first century A.D., is best known for his dialogues, letters, and tragedies, but he is also of interest in the history of science for his *Natural Questions*. This work deals mostly with topics in physics, meteorology, and astronomy, where his sources are principally Aristotle and Theophrastus. He writes of the low state to which general knowledge of astronomy had fallen since the days when the ancient Greeks "counted the stars and named every one," noting that "there are many nations at the present hour who merely know the face of the sky and do not understand why the moon is obscured in an eclipse." But he is hopeful that "the day will come when the progress of research through the long ages will reveal to sight the mysteries of nature that are now concealed."

During the early medieval period the attitude of Christian scholars was that the study of science was not necessary, for in order to save one's soul it was enough to believe in God, as Saint Augustine of Hippo (354–430) wrote in his *Enchiridion*: "It is enough for Christians to believe that the only cause of created things, whether heavenly or earthly, whether visible or invisible, is the goodness of the Creator, the one true God, and that nothing exists but Himself that does not derive its existence from Him."

The Greek philosopher Plotinus (205–70) is believed to have been born in Egypt; he studied at Alexandria before moving to Rome at the age of forty. His works cover the whole range of philosophy, including cosmology and physics. They embody a synthesis of Platonic, Pythagorean, Aristotelian, and Stoic thought that came to be known as Neoplatonism, the dominant philosophy in the Greco-Roman world through the rest of antiquity and on into the medieval era.

Iamblichus (250–c. 325), a scholar of Syrian origin, studied in Rome under the Neoplatonist philosopher Porphyry and later established his own school in Syria. His extant work consists of nine books on the

Pythagoreans, including their number mysticism, as well as books on the use of arithmetic in physics, ethics, and theology. Iamblichus goes far beyond Plato in advocating the complete mathematization of nature, for he felt that mathematics was the key to understanding not only the movement of the celestial bodies, but terrestial phenomena as well.

The Roman writer Chalcidius, who flourished in the fourth century, is noted for his Latin translation of Plato's *Timaeus*, as well as for his own commentary on that work. These were the only sources of knowledge about Plato's cosmology available in Europe during the early medieval era. Chalcidius was also influenced by the ideas of Aristotle and transmitted them to medieval Europe as well, though in somewhat modified form. Two of the most important ideas of Aristotelian science that were thus perpetuated were the concept of the four elements and the astronomical theory of the homocentric spheres, which included the notion of the dichotomy between the terrestrial and celestial regions. Chalcidius refers to the astronomical theories of Heraclides Ponticus, who is also mentioned by the Roman Neoplatonists Macrobius and Martianus Capella.

Macrobius, who may have been from North Africa, flourished in the early fifth century. Besides his mention of the theories of Heraclides, Macrobius also writes of the number mysticism of the Pythagoreans and says that several Platonists believed that the interplanetary distances were such as to produce harmonious relations, the famous "harmony of the spheres."

Martianus Capella was of North African origin and flourished in the years 410–39. He is the author of an allegorical work on the seven liberal arts entitled *The Nuptials of Mercury and Philology*. In Book VIII of this work, an introduction to astronomy, he states that Mercury and Venus orbit the sun, a theory that he attributes to Heraclides Ponticus, though probably mistakenly. His work was very popular in the early medieval era in Europe, when a number of commentaries were written on it.

Proclus (ca. 410–85), the last great Neoplatonist philosopher, was born in Constantinople but moved to Athens in his youth to study at the Academy. He remained in Athens for the rest of his days, except for a brief exile due to his paganism, and was head of the Academy during his latter years. He was the last great synthesizer of Greek philosophy and as such was extremely influential in medieval and Renaissance thought.

His works include a commentary on Book I of Euclid's *Elements* that contains a rich history of Greek geometry, and a treatise entitled *Outline of Astronomical Hypotheses*, a summary of the theories of Hipparchus and Ptolemy.

The Roman scholar Boethius (ca. 480–524) held high office under the Ostrogoth king Theodoric, who had him imprisoned and executed. His best-known work is his *Consolation of Philosophy*, written while he was in prison before his execution. Boethius's other works fall into two categories: his translations from Greek into Latin of Aristotle's logical works, and his own writings on logic, theology, music, geometry, and arithmetic. His writings played a substantial role in the transmission to medieval Europe of the basic parts of Aristotle's logic and of elementary arithmetic.

Cassiodorus was another Roman who held high office under the Ostrogoths. In his *Introduction to Divine and Human Readings* he urged the monks of the monastery at Monte Cassino, in Italy, founded by Saint Benedict in 529, to copy faithfully the classics of ancient scholarship preserved in their libraries as a cultural heritage. He listed some of the important works of science that he thought should be preserved, and in doing so he described and thus transmitted the basic Aristotelian classification of the sciences. This classification divides philosophy into theoretical and practical areas. The theoretical areas are metaphysics, physics, and mathematics, the latter being further subdivided into arithmetic, music, geometry, and astronomy; the practical areas are ethics, economics, and politics. The mathematical sections of the book are briefer and more elementary, mostly dealing with definitions. There is also a section on medicine, in which Cassiodorus gives advice on the use of medicinal herbs, urging the monks to read the works of Hippocrates, Dioscorides, and Galen.

Isidore of Seville (560–636), a Visigothic bishop, wrote the first European encyclopedia, the *Etymologies*, which incorporated compilations of all the Roman authors to whom he had access. Despite the very low scientific level of this work, it achieved wide popularity in medieval Europe as a source of knowledge of all kinds from astronomy to medicine.

The Venerable Bede (674–735), an English monk, is noted for his *De Rerum Natura*, based largely on the *Etymologies*, to which he added Pliny's *Natural History*, a work that had not been available to Isidore. Although derivative, *De Rerum Natura* is greatly superior to the *Etymologies* and the

Natural History because Bede's approach is much more critical than those of Isidore and Pliny. This is evident as well in Bede's *De Temporum Ratione*, which contains his researches on the solar and lunar cycles that he used to calculate future dates for the celebration of Easter, and that also allowed him to explain tidal action as being due to the influence of the sun and moon. Bede's two books were remarkable achievements for their time, and for centuries afterward they were Europe's principal sources of knowledge concerning nature.

Bede's work forms part of a tradition, established in English and Irish monasteries at a very early date, in which the study of science took an honored place along with theological studies, such as in the schools founded as early as the sixth century at Clonard, Bangor, and Iona. This tradition was greatly stimulated when two Greek-speaking monks— Theodore of Tarsus and Adrianus Africanus—were sent to England by Pope Vitalianus (r. 657–68), who thus instituted the study of Greek language and culture in northern Europe.

This cultural revival resulted in the founding of cathedral schools both in England and on the Continent. Alcuin (735–804) was a particularly important figure in the spread of this revival, moving from the cathedral school at York to the court of King Pepin the Short, where he was influential in sparking the Carolingian Renaissance.

Meanwhile, the center of the Roman Empire had moved eastward. This shift had been made official in 330, when Constantine the Great transferred his capital from Italy to the Greek city of Byzantium on the Bosphorus, the "New Rome," which then came to be called Constantinople. Constantine was baptized just before he died, seven years later, the first step in establishing Christianity as the state religion of the empire, a process that was virtually completed by the second half of the following century.

During that period, imperial sovereignty was sometimes divided between emperors of the western and the eastern halves of the territory, the latter ruling in Constantinople. The third quarter of the fifth century was a chaotic period in the history of the western part of the empire. Ten men followed in turn as emperor of the western empire, the last being Romulus Augustus, who was overthrown in 476. By that time most of western Europe had been lost to the empire, and thenceforth the emperor in Constantinople was sole ruler of what remained.

Constantine had founded a university in Constantinople, and this institution was reorganized in 425 by Theodosius II. The new university, which became the most important center of learning in the empire, originally had twenty chairs of grammar, equally divided between Greek and Latin, and eight in rhetoric, five of which were in Greek and three in Latin, as well as two professorships in law and one in philosophy. By the following century Latin had fallen out of use in Constantinople and all of the professorships at the university were in Greek. This was part of the great cultural divide that developed in the early medieval era between the Latin west and the Greek east, a dichotomy that separated the newly emerging civilization of western Europe from the Byzantine world of the Balkans and Asia Minor.

Constantine had organized the first ecumenical council of the church, which had been held in 325 at Nicaea. The second ecumenical council was held at Constantinople in 381, the third at Ephesus in 431, and the fourth in 451 at Chalcedon, in the Asian suburbs of the capital, the principal business at all of these synods being doctrinal matters, particularly concerning the nature of Christ. The bishops at Chalcedon formulated what became the orthodox Christological doctrine: that Christ was both human and divine, his two natures being perfect and indivisible though separate. At the same time they condemned as heretics those who thought differently, namely the Monophysites, who believed that Christ had a single nature, and the Nestorians, who thought that he had a dual nature. The Monophysites and Nestorians, whose believers were principally in southeastern Asia Minor, Syria, Mesopotamia, Persia, and Egypt, then formed their own schismatic churches.

The Nestorians had already founded important schools in northern Mesopotamia at Edessa (Turkish Urfa) and Nisibis, where the language of instruction was Syriac, a Semitic language deriving from Aramaic. Among the books used at these schools were Greek treatises translated into Syriac, most notably the logical works of Aristotle. The school at Edessa was closed by the emperor Zeno in 489, whereupon the Nestorian scholars moved east to Nisibis, which was then in Persian territory.

The eastward migration of Nestorians eventually brought them to the Sassanid capital at Jundishapur, in western Persia, where in the late fifth century they joined the faculty of a medical school that had been founded by King Shapur I (r. 241–72). There the Nestorian faculty taught Greek philosophy, medicine, and science in Syriac translations. The school became a center for the translation of works in medicine, cosmology, astronomy, and Aristotelian philosophy; the languages involved at various times included Greek, Syriac, Sanskrit, Pahlavi, and, subsequently, Arabic.

The best of the early Syriac translators was Sergius of Reshaina (d. 536), a Monophysite priest and physician who had been educated in the Platonic school of Ammonius in Alexandria. His translations from Greek into Syriac included some of Aristotle's logical works, which were at about the same time being rendered from Greek into Latin by Boethius. He also wrote two works of his own on astronomy, *On the Influence of the Moon* and *The Movement of the Sun*, both undoubtedly based on Greek sources. Sergius was characterized by a later Syriac writer as "a man eloquent and greatly skilled in the books of the Greeks and Syrians and a most learned physician of men's bodies. He was orthodox in his opinions . . . but his morals [were] corrupt, depraved and stained with lust and avarice."

The most distinguished Syriac scholar of the early medieval period was Severus Sebokht (d. 667), a Nestorian bishop who wrote on both scientific and theological subjects. His scientific writings included works on logic and astronomy as well as the earliest known treatise on the astrolabe, an astronomical instrument almost certainly of Hellenistic origin. He was also the first to use the so-called Hindu-Arabic number system. Writing in 662, he praised the Hindus and "their valuable methods of calculation, and their computation that surpasses description." He went on to remark, "I only wish to say that this computation is done by means of nine signs," indicating that the symbol for zero had not yet made its appearance.

By the sixth century the character of the Roman Empire had changed profoundly from what it had been in the time of Augustus, Greek having replaced Latin as the dominant language in the capital and Christianity triumphant over the old Greco-Roman gods. Modern historians consider the sixth century to be a watershed in the history of the empire, which from that time on they tend to call Byzantine rather than Roman. As the great churchman Gennadius was to say in the mid-fifteenth century, in the last days of the Byzantine Empire: "Though I am a Hellene by speech yet I would never say that I was a Hellene, for I do not believe as Hellenes believed. I should like to take my name from my faith, and if anyone asks me what I am I answer, 'A Christian.' Though my father dwelt in Thessaly I do not call myself a Thessalian, but a Byzantine, for I am of Byzantium."

The peak of the Byzantine Empire came under Justinian I (r. 527–65), who reconquered many of the lost dominions of the empire, so that the Mediterranean once again became a Roman sea. Justinian also broke the last direct link with the classical past when, in 529, he issued an edict forbidding pagans to teach. As a result the ancient Platonic Academy in Athens was closed, ending an existence of more than nine centuries, as its teachers went into retirement or exile.

Those who went into exile included Damascius, the last director of the Academy, along with Isidorus of Miletus, who had been his predecessor, and Simplicius of Cilicia. They and four other scholars from the Academy were given refuge in 531 by the Persian king Chosroes I (r. 531–79), who appointed them to the faculty of the medical school at Jundishapur. The following year the seven of them were allowed to come back from their exile, six of them returning to Athens, while Isidorus took up residence in Constantinople.

Simplicius is renowned for his commentaries on Aristotle, which contain much valuable material otherwise unavailable, including fragments of the pre-Socratic philosophers. He was a staunch supporter of Aristotle's, some of whose ideas were criticized by John Philoponus, who had succeeded Ammonius as head of the Platonic school in Alexandria. Thus in the twilight of antiquity a great debate took place about the Aristotelian worldview, which was attacked by John Philoponus and defended by Simplicius. The most interesting part of this debate focused on why a projectile, such as an arrow, continues moving after it receives its initial impetus. Philoponus rejected the Aristotelian theory presented by Simplicius, which was that the air displaced by the arrow flows back to push it from behind, an effect called *antiperistasis*. Instead Philoponus suggested that the arrow, when fired, receives an "incorporeal motive force," an idea that was revived in Persia by Ibn Sina and later, in medieval Europe, where it was known as the "impetus theory."

Justinian appointed Isidorus to be chief of the imperial architects, along with Anthemius of Tralles, their task being to design and build the great church of Haghia Sophia in Constantinople, whose foundation was laid in 532. Anthemius died during the first year of construction, but Isidorus carried the work through to completion, after which Justinian dedicated the church, on 26 December 537. Haghia Sophia, which some consider to be the greatest building in the world, still stands today, a symbol of the golden age of the Byzantine Empire under Justinian.

Isidorus and Anthemius studied and taught the works of Archimedes and the Archimedean commentaries of Eutocius of Ascalon. Isidorus was apparently responsible for the first collected edition of at least the three Archimedean works commented upon by Eutocius—*On the Sphere and Cyclinder, On the Measurement of the Circle,* and *On the Equilibrium of Planes*—as well as the commentaries themselves.

Later Byzantine writers added other works to this collection, most notably Leo the Mathematician (ca. 790–ca. 869). Leo's collection of the writings of Archimedes included all of the works now known, excepting *On Floating Problems, On the Method, Stomachion,* and *The Cattle Problem.* Leo also transcribed Ptolemy's *Almagest* as well as the collection of Ptolemy's mathematical and astronomical writings known as the *Little Astronomy.*

One of Leo's students was captured by an Arab army in 830 and ended up at the court of the Abbasid caliph al-Ma'mun (r. 813–33). The caliph, who was sponsoring translations of ancient Greek science and mathematics into Arabic, thus learned of Leo's accomplishments and invited him to Baghdad. But the Byzantine emperor Theophilus (r. 829–42) kept Leo in Constantinople by appointing him as the head of a new school of philosophy and science, where he had his students copy manuscripts of Archimedes and Euclid.

Thus the diaspora of classical learning brought Greek thought from Athens to Rome, Constantinople, and Jundishapur. It took root in the new civilizations that emerged in western Europe, Byzantium, and the Islamic world, three streams of culture whose eventual confluence would produce a renaissance of science.

BAGHDAD'S HOUSE OF
WISDOM: GREEK INTO ARABIC

The Islamic era began in 622 with the Prophet Muhammad's flight from Mecca to Medina. Led by the first caliphs, Arab forces conquered Syria in 637, Egypt in 639, Persia in 640, Tripolitania in 647, and northwest Africa in 670. An Arab fleet attacked Constantinople in 674, subjecting the city to an unsuccessful four-year siege. Within the next half century the armies of Islam conquered Sind (a province of western Pakistan) and Transoxiana (in central Asia) and invaded Spain, raiding across the Pyrenees until they were finally stopped by Charles Martel in 732 at the Battle of Tours, in France.

Mu'awiya, leader of the Umayyad clan, assumed the caliphate at Jerusalem in 661, whereupon he immediately moved to Damascus. This was the beginning of the Umayyad dynasty, whose caliphs ruled mostly from Damascus. In 749, after a four-year civil war, the Umayyads were overthrown and Abu'l-Abbas al-Saffah became caliph, thus founding the Abbasid dynasty, which would last for more than five centuries. Abu'l-Abbas was succeeded in 754 by his brother Abu Jafar al-Mansur, who in the years 762–65 built Baghdad as his new capital, beginning the most illustrious period in the history of Islam.

Baghdad emerged as a great cultural center under al-Mansur (r. 754–75) and three generations of his successors, particularly al-Mahdi (r. 775–85), Harun al-Rashid (r. 786–809), and al-Ma'mun (r. 813–33). According to the historian al-Mas'udi (d. 956), al-Mansur "was the first caliph to have books translated from foreign languages into Arabic," including "books by Aristotle on logic and other subjects, and other ancient books from classical Greek, Byzantine Greek, Pahlavi, Neo-Persian, and Syriac." Al-Mas'udi says that al-Mansur "was the first caliph to favor astrologers and to act on the basis of astrological prognostications." He notes that the caliph had three astronomers in his court who also served as astrologers: Nawbaht the Zoroastrian, Ibrahim al-Fazri, and Ali ibn Isa the Astrolabist.

Nawbaht was the first court astrologer of al-Mansur, and his examination of the celestial signs led him to advise the caliph to begin the construction of Baghdad on 30 July 762. Ibrahim al-Fazari, the first court astronomer of the Abbasid caliphs, worked on problems of calendar reform. Ali ibn Isa is credited with being the first in Islam to make an astrolabe, an ancient Greek astronomical instrument much used by Islamic astronomers. He was also a physician, famous for his *Notebook of the Oculists*, the earliest important Islamic treatise on the structure and illnesses of the eye.

Nawbaht was succeeded as court astrologer by his son Abu Sahl ibn

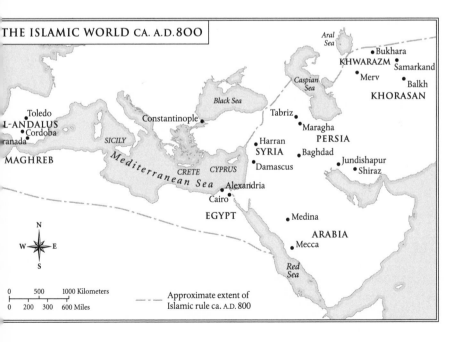

Nawbaht, the author of *Kitab an Nahmutan*. This was the first book in Arabic on astrological history, a dynastic chronicle in terms of cyclical periods of varying lengths governed by celestial bodies. He writes, "The people of every age acquire fresh experience and have knowledge renewed for them in accordance with the decree of the stars and the signs of the zodiac, a decree which is in charge of governing time by the command of God Almighty." Abu Sahl's motive was to show that the Abbasid succession was preordained by the stars and God, and that it was now their dynasty's turn to renew knowledge.

Theophilus of Edessa (d. 789), a Nestorian Christian who was al-Mahdi's court astrologer, called astrology the "mistress of all sciences," because of the importance of astrological history to the Abbasids and the commissioning of horoscopes by the caliphs. Ptolemy's *Tetrabiblos*, the foremost astrological work of antiquity, was translated from Greek to Arabic by the Christian scholar al-Bitriq during the reign of al-Mansur.

The most prominent astrologer in the early Abbasid period was Mash'allah, a Jew from Basra who was one of those whose examinations of the celestial signs led to the founding of Baghdad. His horoscopes can be dated to the period 762–809, and he served as astrologer to all of the caliphs from al-Mansur to al-Ma'mun. He wrote on every aspect of astrology, his most interesting work being his astrological history. His works were translated into Latin, and he is referred to by Copernicus.

Jabir ibn Hayyan (ca. 721–ca. 815), known in the West as Geber the Wise, is renowned in the Muslim world as the founder of alchemy. He was born in Kufa in southern Mesopotamia and in his later years moved to Baghdad, where he became the court astrologer of Harun al-Rashid. The Gabiran corpus represents virtually all that is known of alchemy in Islam during the early Abbasid period. A basic concept of Islamic alchemy that had been inherited from the ancient Greeks was the notion that materials like sulphur and mercury could be transmuted into gold. Islamic alchemy also involved astrology, astral cosmology, magic, and other occult sciences. These branches of learning came under the heading of the "hidden" (*khafiyyah*) sciences, in contrast to the "open" (*jaliyyah*) sciences such as mathematics.

Another reason for translating Greek works into Arabic was to educate the secretaries needed to administer the Abbasid empire. This is evident in the writings of Ibn Qutayb (d. 889), whose *Adab al-Katib* (Education of the Secretaries) enumerates the subjects that state secretaries should learn, including arithmetic, geometry, and astronomy, as well as practical skills such as surveying, metrology, and civil engineering. It was Greek science that could provide the necessary textbooks.

Al-Mansur suffered from dyspepsia, or chronic indigestion, and soon after he moved into his new capital he sought the aid of the physicians at the Jundishapur medical school. His ailment was cured by the director of the hospital, Gurgis ibn Buhtisu, a Nestorian Christian, who came to Baghdad to serve as al-Mansur's personal physician. The Buhtisus became the leading practitioners of medicine in Baghdad, several generations of them serving as personal physicians to the caliphs. The historian Ibn Usaybi'a reports that al-Mansur commissioned many translations of Greek works from Gurgis ibn Buhtisu. The translations would have been done from Syriac by Nestorian scholars from Jundishapur, whose medical center was eventually transferred to Baghdad, becoming the first hospital and Islamic school of medicine in the Abbasid capital.

The translation program continued under al-Mansur's son and successor al-Mahdi. The new caliph's vizier, Yahya ibn Khalid ibn Barmak, is credited by the tenth-century Tunisian scholar Abdallah ibn abi-Zayd with initiating the Abbasids' policy of importing Greek books from the Byzantine Empire.

The most important intellectual institution in Baghdad under the Abbasids was the famous Bayt al-Hikma, or House of Wisdom, which originally seems to have been basically a library. Pahlavi manuscripts were kept there and in the early Abbasid period some of these were translated into Arabic. The chronicler Ibn al-Nadim notes that the court astrologer Abu Sahl ibn Nawbaht was employed by Harun al-Rashid at the Bayt al-Hikma, where "he translated from Persian into Arabic and relied in his scholarship on the books of Iran." During the reign of al-Ma'mun astronomers and mathematicians were associated with the Bayt al-Hikma, which at that time probably served as a research institute as well as a library. Ibn al-Nadim also reports that the famous mathematician and astronomer Muhammad ibn Musa al-Khwarizmi (fl. ca. 828) "was employed full-time in the *Bayt al-Hikma* in the service of al-Ma'mun."

Al-Khwarizmi is renowned for his treatise *Kitab al-Jabr wa'l-Muqabalah*, known more simply as *Algebra*, for it was from this work that Europe later learned the branch of mathematics known by that name. In his preface the author writes that the caliph al-Ma'mun "encouraged me to compose a compendious work on algebra, confining it to the fine and important parts of its calculation, such as people constantly require in cases of inheritance, legacies, partition, law-suits and trade, and in all their dealings with one another, or where surveying, the digging of canals, geometrical computation, and other objects of various sorts and kinds are concerned."

Another of al-Khwarizmi's mathematical works survives only in a unique copy of a Latin translation entitled *De Numero Indorum* (Concerning the Hindu Art of Reckoning), the original Arabic version having been lost. This work, probably based on an Arabic translation of works by the Indian mathematician Brahmagupta (fl. 628), describes the Hindu numerals that eventually became the digits used in the modern Western world. The new notation came to be known as that of al-Khwarizmi, corrupted to "algorism" or "algorithm," which now means a procedure for solving a mathematical problem in a finite number of steps that often involves repetition of an operation.

Al-Khwarizmi is the author of the earliest extant original work of Islamic astronomy, the *Zij al-Sindhind* (a *zij* is an astronomical handbook with tables). This is a set of planetary tables using earlier Indian and Greek astronomical elements, including the epicycle theory. He and Fadil ibn al-Nawbaht are credited with building the first Islamic observatory, which they founded in Baghdad circa 828, during the reign of al-Ma'mun. Al-Khwarizmi also wrote the first comprehensive Islamic treatise on geography, in which he revised much of Ptolemy's work on this subject, drawing new maps.

Euclid's *Elements* was first translated into Arabic during the reign of Harun al-Rashid by the mathematician al-Hajjaj ibn Matar (fl. ca. 786–833), under the patronage of the vezir Yahya ibn Khalid ibn Barmak. Al-Hajjaj did an improved and abbreviated version of the *Elements* for al-Ma'mun, apparently for use as a school textbook.

Al-Mahdi commissioned the translation of Aristotle's *Topics* into Arabic from Syriac, into which it had been translated from Greek. Later the work was translated directly from Greek into Arabic. The motivation for translating the *Topics* was that it taught the art of systematic argumentation, which was vital in discourse between Muslim scholars and those of other faiths and in converting nonbelievers to Islam, which became state policy under the Abbasids. Aristotle's *Physics* was first translated into Arabic during the reign of Harun al-Rashid, the motivation apparently being its use in theological disputations concerning cosmology.

Harun al-Rashid's Baghdad is the setting for the *Thousand Nights and One Night*, where the "Tale of Aladdin and His Wonderful Lamp" reflects the marvels of the new science and the amazing inventions and discoveries that were attributed to wizard scientists. The medieval Islamic scientist was, at least in the popular view, the prototype of the villainous Moor who led Aladdin to the lamp. As Shaharzad says of the Moor: "From his earliest youth he had studied sorcery and spells, geomancy and alchemy, astrology, fumigation and enchantment; so that after thirty years of wizardry, he had learnt the existence of a powerful lamp in some unknown place, powerful enough to raise its owner above the kings and powers of the world."

Harun al-Rashid's son al-Ma'mun continued the translation program of his predecessors. Al-Mas'udi writes of al-Ma'mun's interest in astrology and his patronage of intellectual investigations: "At the beginning of his reign . . . he used to spend time investigating astrological rulings and prognostications, to follow what the stars prescribed, and to model his conduct on that of the past Sassanian emperors. . . . He had jurists and the learned among men of general culture attend his session; he had such men brought from various cities and stipends for them allocated."

Islamic astronomy was dominated by Ptolemy, whose works were translated into Arabic and also disseminated in summaries and commentaries. The earliest Arabic translation of the *Almagest* is by al-Hajjaj ibn Matar in the first half of the ninth century. The most popular compendium of Ptolemaic astronomy was that of al-Farghani (d. after 861), who used the findings of earlier Islamic astronomers to correct the *Almagest*. Habash al-Hasib (d. ca. 870) produced a set of astronomical tables in which he introduced the trigonometric functions of the sine, cosine, and tangent, which do not appear in Ptolemy's works.

Islamic science developed apace with the translation movement, which involved philosophers as well as scientists. The founding of Islamic philosophy is credited to Abu Yusuf Yaqub ibn Ishaq al-Kindi (ca. 801–866), the Latin Alkindes, famous in the West as the "Philosopher of the Arabs." Al-Kindi was from a wealthy Arab family in Kufa, in present-day Iraq, which he left to study in Baghdad. There he worked in the Bayt al-Hikma, enjoying the patronage of al-Ma'mun and his immediate successors.

Al-Kindi, though not a translator himself, benefited from the translation movement to become the first of the Islamic philosopher-scientists, founding the Aristotelian movement in Islam. He was a polymath, his treatises including works in geography, politics, philosophy, cosmology, physics, mathematics, meteorology, music, optics, theology, alchemy, and astrology. He was the first Islamic theorist of music, following in the Pythagorean tradition. His work on optics follows Theon of Alexandria in studying the propagation of light and the formation of shadows, and his theory of the emission and transmission of light is based on that of Euclid. Al-Kindi's ideas on visual perception, which differed from those of Aristotle, together with his studies of the reflection of light, laid the foundations for what became, in the European Renaissance, the laws of perspective. His studies of natural science convinced him of the value of rational thought, and as a result he was the first noted Islamic philosopher to be attacked by fundamentalist Muslim clerics. His *Letter on the Method of Banishing Sadness* says that the cure for melancholia is applying oneself to the only enduring object, the world of the intellect.

Al-Kindi also wrote a work called *The Theory of the Magic Art, or On Stellar Rays,* which survives only in medieval Latin manuscripts. He begins the treatise by saying that stellar rays are emitted by celestial bodies and influence everything in the universe, mankind included, and that a study of the heavens thus allows astrologers to predict the future. He concludes with a discussion of the magical power of talismanic inscriptions, an occult art that is still practiced in Islam. "The sages," he writes, "have proved by frequent experiments that figures and characters inscribed by the hand of man on various materials with intention and due solemnity of place and time and other circumstances have the effect of motion upon eternal objects."

The most important figures in the program for sponsoring science under al-Ma'mun and his immediate successors were the Banu Musa, three brothers named Muhammad, Ahmad, and al-Hasan. These were the sons of Musa ibn Shakir, a former highway robber who became an astrologer in Merv, where he befriended al-Ma'mun before the latter became caliph in 813. When Musa died his three sons were adopted by al-Ma'mun, who had them educated in Baghdad after he became caliph. After finishing their studies the Banu Musa served al-Ma'mun and his immediate successors in various ways, becoming rich and powerful in the process. They spent much of their wealth in collecting ancient manuscripts, and they also supported a group of translators in Baghdad. The Banu Musa themselves are credited with writing some twenty books on astronomy, mathematics, and engineering, three of which have survived, including a treatise by Ahmad on ingenious mechanical devices patterned on the *automata* of Hellenistic Greece.

Ibn Khallikan also tells the story of how al-Ma'mun directed the Banu Musa to measure the circumference of the earth, to verify the measurements made by Eratosthenes and other ancient Greek scientists. The method used by the Banu Musa was to measure the north-south distance between two points in the Sinjar desert where the elevation of the polestar differed by one degree, whereupon they multiplied this by 360 to obtain the circumference of the earth. They value they obtained, according to Ibn Khallikan, was 8,000 *farsakhs*, or 24,000 miles; the presently accepted value is 24,092 miles.

The two most famous translators of the period in Baghdad were Hunayn ibn Ishaq and Thabit ibn Qurra, both of whom were employed by the Banu Musa "for full-time translation," according to the chronicler Abu-Sulayman al-Sigistani, who says they were paid a salary that put them on a par with the highest officials in the government bureaucracy.

Hunayn ibn Ishaq (808–73), known in Latin as Jannitus, was born at al-Hira in southern Iraq, the son of a Nestorian apothecary. He went to Baghdad to study under the Nestorian physician Yuhannah ibn Masawayh (d. 857), the personal physician to al Ma'mun and his successors. Hunayn, who at the time knew only Syriac, was disappointed by Ibn Masawayh, who discouraged him when he inquired about Greek medical texts. According to his autobiography, the *Risala*, Hunayn then went away to "the land of the Greeks" (*bilal-al-Rum*, probably Byzantium) and obtained a sound knowledge of Greek, after which he lived in Basra for a time to learn Arabic. He then moved to Baghdad, where he and his students, who included his son Ishaq ibn Hunayn and his nephew

Hubaysh, made meticulous translations from Greek into both Syriac and Arabic. Their translations included the medical works of Hippocrates and Galen, Euclid's *Elements*, and Dioscorides' *De Materia Medica*, which became the basis for Islamic pharmacology. Ishaq's extant translation of Aristotle's *Physics* is the last and best version of that work in Arabic. His translations included Ptolemy's *Almagest*, while his father, Hunayn, revised the *Tetrabiblos*. Hunayn also revised an earlier translation of Galen by Yahya ibn al-Bitriq (d. 820); these were synopses that contained Plato's *Republic, Timaeus,* and *Laws,* the first rendering of the Platonic dialogues into Arabic.

Hunayn was indefatigable in his search for Greek manuscripts, as he writes in regard to Galen's *De Demonstratione:* "I sought for it earnestly and traveled in search of it in the lands of Syria, Palestine, and Egypt until I reached Alexandria, but I was not able to find anything except about half of it in Damascus."

Hunayn was an outstanding physician and wrote two books on medicine, both extant in Arabic, one of them a history of the subject, the other a treatise entitled *On the Properties of Nutrition,* based on Galen and other Greek writers. His other works include treatises on philosophy, astronomy, mathematics, optics, ophthalmology, meteorology, alchemy, and magic, and he is also credited with establishing the technical vocabulary of Islamic science.

One of the important translators and scientists, Thabit ibn Qurra (ca. 836–901), was born in the Mesopotamian town of Harran, a center of the ancient Sabean cult, an astral religion in which the sun, moon, and five planets were worshipped as divinities. Harran had preserved Hellenic literary culture, and so educated Sabeans like Thabit were fluent in Greek as well as in Syriac and Arabic. Thabit was working as a money changer in Harran when he was "discovered" by Muhammad ibn Musa, one of the Banu Musa brothers, who was returning from an expedition to find ancient Greek manuscripts in the Byzantine Empire. Muhammed brought the young Thabit back with him to Baghdad, where he became one of the salaried translators who worked for the Banu Musa along with Ishaq ibn Hunayn. After Thabit established himself, a number of his fellow Sabeans joined him in Baghdad, where they formed a school of mathematics, astronomy, and astrology that lasted through three generations.

Thabit translated works from both Syriac and Greek into Arabic, including the *Introduction to Arithmetic* by Nichomachus as well as improved editions of Euclid's *Elements* and Ptolemy's *Almagest*. His descendants also produced Arabic translations, in particular, renderings of the writings of Archimedes and Apollonius of Perge.

Thabit wrote treatises of his own as well, including works on physics, astronomy, astrology, dynamics, mechanics, optics, and mathematics. He wrote a commentary on Aristotle's *Physics* and an original work entitled *The Nature and Influence of the Stars*, which laid out the ideological foundations of Islamic astrology. He also wrote a comprehensive work on the construction and theory of sundials.

Thabit revived the erroneous "trepidation theory" of Theon of Alexandria. This held that the pole of the heavens oscillated back and forth, as opposed to the correct theory, first given by Hipparchus, that the celestial pole precessed in a circular path. Thabit pictured the planets as being embedded in solid spheres with a compressible fluid between the orbs and the eccentric circles. His planetary theory included a mathematical analysis of motion, in which he referred to the speed of a moving body at a particular point in space and time, a concept that is now part of modern kinematics. His contributions in mathematics include calculating the volume of a paraboloid and giving geometrical solutions to some quadratic and cubic equations. He also formulated a remarkable theorem concerning so-called amicable numbers, where each number of an "amicable" pair is the sum of the proper divisors (all of its factors except itself) of the other, the smallest such pair being 220 and 284.

Another prominent figure of the translation movement was Qusta ibn Luqa, a Greek-speaking Christian from the Lebanon, who worked in Baghdad as a physician, scientist, and translator until his death in 913. His translations included works by Aristarchus, Hero, and Diophantus. He wrote commentaries on Euclid's *Elements* and Dioscorides' *De Materia Medica*, as well as original treatises on medicine, astronomy, metrology, and optics. His medical works include a treatise on sexual hygiene and a book on medicine for pilgrims.

Qusta also wrote a work on magic entitled *Epistle Concerning Incantations, Adjurations and Amulets,* a Latin translation of which was cited by Albertus Magnus in the thirteenth century. Qusta's attitude toward sorcery is evident from an anecdote in this book, where he tells the story of "a certain great noble of our country" who believed that a witch had made him impotent. Qusta advised the noble to rub himself down with the gall of a crow mixed with sesame, persuading him that this was an aphrodisiac, and this gave the man such confidence that he overcame his imaginary ailment and regained his sexual powers.

The program of translation continued until the mid-eleventh century, both in the East and in Muslim Spain. By that time most of the important works of Greek science and philosophy were available in Arabic translations, along with commentaries on these works and original treatises by Islamic scientists that had been produced in the interim. Thus, through their contact with surrounding cultures, scholars writing in Arabic were in a position to take the lead in science and philosophy, absorbing what they had learned from the Greeks and adding to it to begin an Islamic renaissance, whose fruits were eventually passed on to western Europe. Islamic scholars dated the beginning of this age of enlightenment to the reign of Harun al-Rashid, as in the encomium written by the poet Mosuli:

Did you not see how the sun came out of hiding on Harun's accession and flooded the world with light?

6

THE ISLAMIC RENAISSANCE

The Islamic renaissance began even before the translation movement ended, spreading eastward to central Asia and westward into North Africa and Spain, giving rise to works in all of the branches of science known to the ancient Greeks. The preponderance of the earliest figures in this renaissance worked in the region between Baghdad and central Asia, where Arabic science would continue to flourish long after it had declined elsewhere in the Islamic world, particularly in astronomy.

The rapid expansion of early Islam, together with the requirement that all Muslims make a pilgrimage to Mecca, stimulated an interest in geography and natural history among Arabic scholars. The most popular of the early Islamic works in this area were by Abu'l Hasan al-Masudi, who has been called the Muslim Pliny.

Al-Masudi was born toward the end of the ninth century near Baghdad, where he studied before traveling throughout Asia and parts of Europe, reporting that he found the Europeans "humorless, gross and dull." He spent the last decade of his life in Syria and Egypt, where he died in 956. His best-known work is *Prairies of Gold*, which reveals him as a traveler, chronicler, geographer, geologist, and natural historian. Al-Masudi's last work was his *Book of Indications and Revisions*, which summarizes his observations and philosophy concerning nature and history. He also developed theories of music and advocated musical therapy, as well as proposing the concept of human evolution. He cautioned against an uncritical acceptance of the "ancients," believing that science advanced through new discoveries.

Astronomy always held pride of place among the sciences in Islam, and Arabic astronomers often waxed eloquent in extolling the utility and godliness of their field. Muhammad ibn Jabir al-Battani (858–929) begins his *Zij al-Sabi* by citing a verse of the Koran in praise of astronomy: "He it is who appointed the sun a splendor and the moon a light, and measured for her stages, that ye might know the number of the years, and the reckoning."

Al-Battani, the Latin Albategnius, was a Sabean from Harran who had a private observatory in the Syrian town of Al-Raqqa. His *Zij al-Sabi*, known in its Latin translation as *De Scientia Stellarum* (On the Science of Stars), was used in Europe until the end of the eighteenth century. In the preface to his *Zij*, al-Battani writes that the errors he found in earlier astronomical treatises had led him to improve the Ptolemaic model with new theories and observations, just as Ptolemy had done with the work of Hipparchus and other predecessors. His observations of the changing position of the celestial pole led him to reject the trepidation theory of Thabit ibn Qurra in favor of the ancient precession theory of Hipparchus. Ptolemy had measured the rate of precession to be 1 degree in 100 years, while al-Battani found it to be 1 degree in 66 years; the correct value is 1 degree in 72 years. Al-Battani's astronomical writings were translated into Latin and were used by astronomers until the seventeenth century.

Many Arabic astronomers doubled as astrologers, given the great popularity of astrology among all classes in the Islamic world. This was despite the determined opposition of theologians, who quoted the Koran in reminding the faithful that "none in the heavens and the earth knows the unseen save Allah." The poet Sa'di of Shiraz tells an anecdote in ridiculing the claims that astrology can predict the future. It seems that an astrologer returned home one day and found his wife with another man; when he raised a fuss about this a neighbor mocked him by saying, "What can you know of the celestial sphere when you cannot tell who is in your own house?"

Prominent Arabic philosophers also attacked astrology, the earliest to do so being Abu Nasr al-Farabi (ca. 870–950). Al-Farabi, known in

Latin as Alpharabius, was a Turk born in Transoxiana, the region beyond the Oxus River (the modern Amu Darya) in central Asia, where he spent the first half of his life. He later went to school in Baghdad, where he studied logic under a Syriac-speaking Nestorian scholar named Yuhanna ibn Haylan. After two decades in Baghdad he moved to the court of Saif al-Daulah in Damascus; he lived there for the rest of his days.

Al-Farabi's attack on astrology comes in the introduction to his *Enumeration of the Sciences*, known in its Latin translation as *De Scientiis*. This is the earliest extant medieval classification of the sciences, a system modified and elaborated upon by the Arabic scholars who succeeded him. Despite his opposition to astrology, al-Farabi still includes it with observational and mathematical astronomy as part of the "Science of the Heavens."

Al-Farabi was the second Islamic Aristotelian philosopher-scientist after al-Kindi. But he was also deeply influenced by Plato and made an effort to reconcile Platonic and Aristotelian ideas when they conflicted. His treatise *Al-Madina al-Fadila* (The Good City) attempts to show the relation between Plato's ideal community in the *Republic* and Islam's sharia, or sacred law. He is noted as the Islamic founder of political philosophy and logic. His scientific works include commentaries on Euclid's *Elements* and Ptolemy's *Almagest*. Al-Farabi's treatise on music is the earliest Islamic study of the subject, far ahead of any work in Latin Europe. He was also a composer, and some of his works were played in the rites of the Sufi brotherhoods, a number of them surviving today in the dervish orders of Turkey.

Medicine was another branch of science highly esteemed in Islam, as is evident in one of the hadith, or sayings, attributed to the Prophet Muhammad: "The best gift from Allah is good health. Everyone should reach that goal by preserving it for now and the future."

The first great writer in Islamic medicine is Abu Bakr Muhammad ibn Zakariya al-Razi (ca. 854–ca. 930), the Latin Rhazes, who was born in the Persian city of Rayy. He was famous as a physician in both the East and the West, where he was known as "the Arabic Galen." He studied in Rayy and became the director of the hospital there. He later headed the hospital in Baghdad, and students came from afar to study with him. He is credited with 232 works, of which most are lost, including all of his philosophical treatises. The most important of his surviving medical works is *Al-Hawi*, known in its Latin translation as *Continens*, the longest extant Arabic work on medicine. His treatise on smallpox and measles, known in Latin as *De Peste*, was translated into English and other Western languages and published in forty editions between the fifteenth century and the nineteenth. Al-Razi's writings are characterized by his greater emphasis on observational diagnosis and therapy than on a theory of illnesses and their cures. The titles of some of his works indicate that he was aware of the limitations and misuses of the medical profession, such as his treatises *On the Fact That Even Skillful Physicians Cannot Heal All Diseases* and *Why People Prefer Quacks and Charlatans to Skilled Physicians*.

One of al-Razi's books, known in its English translation as *Spiritual Physick*, concerns the diagnosis and treatment of ailments involving both the body and the soul. Each of the chapters in the book concerns one of twenty psychosomatic diseases; he ends the fourteenth chapter, "On Drunkenness," by quoting an Arabic poem on the evils of drink:

> *When shall it be within thy power*
> *To grasp the good things God doth shower*
> *Though they be but a span from thee,*
> *If all thy nights in revelry*
> *Be passed, and in the morn thy rise*
> *With fumes of drinking in thine eyes*
> *And heavy with its wind, ere noon*
> *Return to thy drunkard's boon?*

Al-Razi's alchemical writings are also well known, particularly the *Book of Secrets*. Here he is less interested in the esoteric philosophical background of alchemy than in the chemical substances, processes, and laboratory equipment involved. Among the substances that he studied was *naft*, or petroleum, which in modern times was to become the principal source of wealth of a number of Islamic countries in the Middle East. He also worked with oil lamps, or *nafata*, for which he used both vegetable oils and refined petroleum as fuel.

Al-Razi wrote on magic and astrology as well as on alchemy, and his work in these fields influenced the first natural philosophers in western

Europe. One of his works, entitled *Of Exorcism, Fascinations, and Incantations*, discusses the use of those occult practices in causing and curing diseases. Those who followed al-Razi's lead searched for the elixir of life, the philosopher's stone, talismans, and the magical properties of plants and minerals with their supposed ability to cure diseases. The most outstanding Islamic physician in the generation after al-Razi was Ali ibn al-Abbas al-Majusi (ca. 925–994), the Latin Haly Abbas. Majusi means "Zoroastrian," although he himself was a Muslim, born near Shiraz. His principal work is *Kitab al-Maliki* (The Royal Book), known in its Latin translation as *Liber Regius*. The main interest of this book today is al-Majusi's assessment of his Greek and Arabic predecessors, including al-Razi.

Al-Majusi emphasized the importance of psychotherapy in treating psychosomatic illnesses; one of those he recognized was unrequited love. His writings on poisons, including their symptoms and antidotes, represents the beginning of medieval toxicology. He wrote on the use of opiates and problems of drug addiction as part of his general discussion of medicines, and he also emphasized chemotherapy. He opposed contraception and the use of drugs to cause abortion, except in cases where the physical or mental health of the woman was endangered. He insisted on the highest standards of medical ethics, referring his colleagues to the Hippocratic code.

The Persian astronomer 'Abd al-Rahman al-Sufi (903–86) was known in the West as Azophi. Little is known of his life and career except his association with the Buyid dynasts, who captured Baghdad in 945 and for more than a century afterward acted as protectors of the Abbasid caliphs, who were reduced to the role of mere puppets. Al-Sufi is best known for his *Treatise on the Constellations of the Fixed Stars*, a critical revision of Ptolemy's star catalog based on his own observations, which was a classic of Arabic astronomy for many centuries and later became known to the West through a Castilian translation. The old Arabic star names that he used were adopted by most later Islamic astronomers and have made their way into modern stellar terminology. The illuminated manuscripts of the *Treatise* are among the most beautiful in Islamic science. The paintings show forty-eight constellations, with tables giving the positions, magnitudes, and colors of all of the stars. Each of the constellations is shown in two facing views: as it would look to an observer on earth, and as it would appear on the celestial sphere to a viewer outside. The mythological figures are portrayed in Islamic rather than Greek costumes, so that in the constellation that bears his name Perseus is dressed in a flowing Arabic robe, brandishing his sword in one hand and with the other holding the severed head of Medusa by her long hair. Some of the most renowned figures of the Islamic renaissance were polymaths who wrote on many different branches of science, including astronomy and not always excluding astrology.

Abu Rayhan al-Biruni (973–1050) is credited with 146 works, comprising treatises on astronomy, astrology, chronology, time measurement,

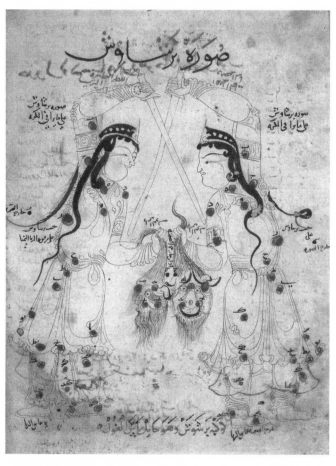

The constellation Perseus shown in al-Sufi's *Treatise on the Constellations of the Fixed Stars*, from a tenth-century Arabic manuscript.

geography, geodesy, cartography, mathematics (including arithmetic, geometry, and trigonometry), mechanics, medicine, pharmacology, meteorology, mineralogy (including gems), history, philosophy, religion, literature, and magic, as well detailed descriptions of his observational instruments and inventions. Probably of Turkish origin, he was born in the central Asian region known as Khwarazm (now in Uzbekistan) and studied under the astronomer and mathematician Abu Nasr Mansur. Later he entered the service of Sultan Mahmud al-Ghaznami of Ghazna (in what is now Afghanistan) in his campaigns of conquest in central Asia and India.

The knowledge that al-Biruni obtained in these campaigns enabled him to write his major work, *The Description of India*, a mine of information on history, geography, science, religion, and studies of humankind and society. This work was also very influential in introducing Hindu mathematics to the Islamic world, which later passed this knowledge on to Europe. His *Chronology of Ancient Nations* describes the calendars and religious festivals of various peoples in antiquity. His *Canon of Al-Mas'ud* became the basic text of Islamic astronomy, just as his *Elements of Astrology* was the standard work in that field. Nevertheless, al-Biruni emphasized that he did not really believe in astrology, for he thought that the "decrees of the stars" had no place in the exact sciences.

Al-Biruni's other accomplishments include an accurate measurement of the earth's circumference; a geared calendar showing the motion of the sun and moon among the signs of the zodiac; a device for making accurate measurements of the specific gravities of liquids; a mechanical triangulation instrument for measuring distances such as the width of a river or the height of a minaret; a mathematical method for determining the qibla, the direction of Mecca from any point; a speculation on the rotation of the earth; and observations on technological processes such as the casting of iron, the production of steel, and the mining and purification of gold, all of the latter techniques and many others described in his *Kitab al-Jamahir*. But al-Biruni's works were never translated into Latin, and so he had little influence on the subsequent development of science in Europe.

Al-Biruni's speculations about celestial motion are extremely interesting, for he disagrees with Aristotle's doctrine of natural place and natural motion, proposing instead that the heavenly bodies do have gravity (i.e., weight) despite the fact that they move in circular orbits rather than toward the center. His thoughts on celestial motion and other matters appear in his correspondence with Abu 'Ali al-Husain ibn Sina, the Latin Avicenna, to whom he addressed a number of questions, the first of which concerns "the possible gravity of the heavens, their circular motion and the denial of the natural place of things."

Ibn Sina (980–1037) was born and educated near Bukhara (in present-day Uzbekistan); he later lived in the Persian towns of Rayy and Hamadan, where he died. He is credited with some 270 works, including an autobiography, completed by his disciple al-Juzjani. His best known works are the *Canon of Medicine* and the *Book of Healing*, which in addition contain chapters on logic, ethics, mathematics, physics, optics, chemistry, biology, botany, geology, mineralogy, meteorology, and seismology. He also wrote on the classification of the sciences, ranking philosophy as "queen of the sciences." His medical writings, along with those of al-Razi, were translated into Latin and used as basic texts in Europe's medical schools until the seventeenth century. His *Canon of Medicine* was far ahead of its times in dealing with such matters as cancer treatment, the influence of the environment, the beneficial effects of physical exercise, and the need for psychotherapy; he recognized the connection between emotional and physical states, including, like al-Majusi, the heartache of unrequited love.

Ibn Sina was the first Muslim scientist to revive the impetus theory of John Philoponus, an attempt to explain why a projectile continues to move after it is fired. He described this impetus as a "borrowed power" given to the projectile by the source of motion, "just as heat is given to water by a fire."

Ibn Sina had immense influence on the subsequent development of science, both in the Islamic world and in Latin Europe, where, as Avicenna, he was known as the "Prince of Physicians." His ideas, which combined Platonic and Aristotelian concepts, had a profound effect on Western thought in the thirteenth century, when a new European science was being created from Greco-Arabic sources.

Ibn Sina's most influential follower was Sayyid Zayn al-Din Ismail al-Juzjani (d. ca. 1070), who lived in the central Asian region of Khwarazm. His principal work is the *Treasury Dedicated to the King of Khwarazm*, a medical encyclopedia based on Ibn Sina's *Canon*, written in Persian, which established the scientific terminology for medicine, including pharmacology. Al-Juzjani's other writings include his *Medical Memoranda* and *The Aims of Medicine*, which, along with his *Treasury*, were the principal sources for the perpetuation of the medical teachings of Ibn Sina and his predecessors. He also wrote a treatise on astronomy, *The Composition of the Heavenly Spheres*, in which he dealt with Ptolemy's controversial concept of the equant, the point about which the planets move with constant velocity, as explained in the *Almagest*, an idea that many Islamic astronomers rejected.

The most illustrious figure in the history of Islamic mathematics is Abu'l Fath Umar ibn Ibrahim al-Khayyami (ca. 1048–ca. 1130), known in the West as Omar Khayyam, the "Tentmaker." Khayyam was born in Nishapur, in Persia, shortly before the Seljuk Turks conquered much of the former Abbasid empire, climaxed by their capture of Baghdad in 1055.

Khayyam's principal work in mathematics is his *Algebra*, which is generally considered to be the culmination of Islamic research in this field, going beyond that of al-Khwarizmi to include cubic equations. He used both arithmetic and geometric methods to solve quadratic equations and employed a scheme of intersecting conics to solve cubic equations, an approach first taken by Archimedes. He was also the first to see the equivalence between algebra and geometry, finally established by Descartes in the seventeenth century.

Khayyam did research in physics and invented a water balance that was for a long time known by his name. He was also involved in a program of calendar reform initiated by the Seljuk sultan Melikshah (r. 1072–92). The Jalali calendar that he and his colleagues created is still used in Iran and other parts of the Islamic world. He refers to this work in one of the quatrains of his *Rubaiyat*, the volume of poetry that made him far more famous in the West than his mathematics.

> *Ah, but my Computations, People say*
> *Reduced the year to better reckoning? Nay*
> *'Twas only striking from the Calendar*
> *Unborn Tomorrow and dead Yesterday*

Islamic theology reached its peak with Abu Hamid al-Ghazali (1058–1111), the Latin Algazel. The most important of al-Ghazali's works is *The Incoherence of the Philosophers*, an attack on the rationalism of Neoplatonist and Aristotelian physics and metaphysics, in which he criticized some of the views of Ibn Sina and al-Farabi. His writings greatly increased the popularity of mysticism in Islam, leading to a rejection of rational philosophy and science. The decline of Arabic science that began in the twelfth century is sometimes attributed partly to the influence of al-Ghazali. Nevertheless, Arabic work in mechanics and astronomy, at least, remained at a high level after his time, particularly in central Asia.

The Archimedean tradition in mechanics and hydrostatics continued to develop in late medieval Islam as far afield as central Asia. This is evidenced by the writings of 'Abd al-Rahman al-Khazini, who in the first half of the twelfth century flourished in Merv, in what is now Turkmenistan. Originally a slave boy of Byzantine origin, possibly a castrato, he seems to have been a high government official under Sanjar ibn Malikshah, who was first emir of Khorasan (r. 1097–1118) and then sultan of the Seljuk Empire (r. 1118–57), during which time Merv became a center of literary and scientific activity.

Al-Khazini's best known work is *The Book of the Balance of Wisdom*, an encyclopedia of medieval mechanics and hydrostatics that also features commentaries on the writings of earlier scholars in this field going back as far as Euclid and Archimedes. The topics covered in the encyclopedia include theories of the center of gravity; measurements of specific gravities of fifty substances, both liquid and solid; determination of the constituents of alloys; and the mechanisms of the steelyard and other balances, including the water balance of Omar Khayyam and one attributed to Archimedes. The encyclopedia also establishes standards of measurement, discusses capillary action, and describes ingenious mechanical *automata*. The most interesting feature of this work is that al-Khazini treats gravitation as a universal force attracting all terrestrial (though not celestial) bodies toward the center of the earth, the attraction being proportional to the weight of the body. He was aware that even air has weight, and that its density decreases with height.

Al-Khazini was also a distinguished astronomer. His most important work in this field was the *Sanjar Zij*, the astronomical tables he compiled for Sultan Sanjar ibn Malikshah, which also includes interesting information on various calendars as well as lists of religious holidays, fasts, rulers, and prophets, concluding with tables of astrological quantities.

Another of his writings on astronomy is the *Treatise on Instruments*. Each of the seven parts of this work is devoted to an astronomical instrument, with instructions for its use as well as explanations of its geometrical basis.

Another Greek scientific tradition that flourished in late medieval Islam was the making of *automata*. Islamic work in this field culminated with the inventions of Badi al-Zaman Abu'l Izz Ismail ibn al-Razzaz al-Jazari (fl. ca. 1200), following in the tradition of Ctesibius, Hero of Alexandria, and Philo of Byzantium.

All that is known of al-Jazari's life is contained in the introduction to his only extant work, the *Book of Knowledge of Ingenious Mechanical Devices*,

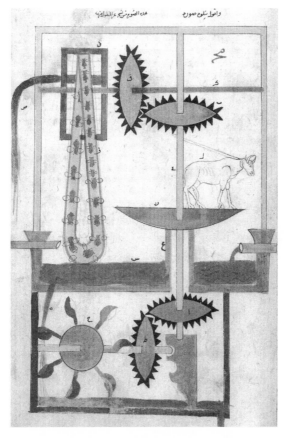

A water-raising machine, from a fourteenth-century Arabic manuscript of al-Jazari's *Book of Knowledge of Ingenious Mechanical Devices*.

which became the definitive text on mechanics and *automata* in the Islamic world. There he says that when he wrote the book he was in the service of Nasr al-Din, the ruler of the Turcoman Artukid emirate that had its capital at Diyarbakir in southeastern Anatolia.

Some of al-Jazari's inventions later reappeared in the West, including his conical valve, mentioned by Leonardo da Vinci. He was renowned for his inventions, some of which had an obvious use, such as pumps and water-raising devices. Others were for decoration or entertainment, including fountains, musical *automata*, and water clocks, while a number were trick vessels of various kinds, all illustrated in miniature paintings.

The Abbasid dynasty came to an end in 1258 with the sack of Baghdad by the Mongols under Hulagu Khan, a grandson of Genghis Khan's. Hulagu had the last Abbasid caliph executed, along with a large part of the population of Baghdad. Whole quarters of the city were looted and destroyed, including the Great Mosque and the Shiite mosque of Khaziman, and chroniclers report that piles of manuscripts were burned, including many that would have come from the House of Wisdom in Baghdad.

Baghdad was never again to be the capital of Islam. It survived only as a provincial city, at the mercy of the successive conquerors who passed that way, until it fell to Tamerlane in 1393. Thenceforth Islamic scholars could only look back on the past glories of the Abbasid capital, as in the encomium the tenth-century geographer Muqaddasi wrote of the city of Harun al-Rashid, referring to "Baghdad, which has no equal in the Orient or in the Western world."

CAIRO AND DAMASCUS

Cairo emerged as an Islamic cultural center soon after its founding in 969 as the new capital of the Fatimids, the dynasty that for the next two centuries ruled North Africa, Egypt, and Syria. Under the caliphs al-Mu'izz (r. 969–75), al-Aziz (r. 975–96), and al-Hakim (r. 996–1021), the Fatimid caliphate in Egypt became the most powerful Islamic state in the world, with Cairo rivaling Baghdad in its splendor.

The enduring symbol of the Fatimid dynasty is the great mosque of al-Azhar, completed in 972 by al-Mu'izz, which became the first Islamic university, still functioning today. The emergence of Cairo as a cultural center was due to al-Hakim, founder of the Dar al-'Ilm, the House of Science, a library whose fame was surpassed only by Baghdad's Bayt al-Hikma, the House of Wisdom. According to the fifteenth-century Egyptian historian al-Maqrizi, the Cairo library had forty rooms. Its collection included eighteen thousand manuscripts dealing with the "Science of the Ancients." Like the ancient Library of Alexandria, the Fatimid Dar al-'Ilm was also a research center and institute of higher learning, its staff including mathematicians, astronomers, and physicians as well as librarians, grammarians, lexicographers, copyists, and readers of the Koran.

The first of the great Islamic scientists to emerge in Fatimid Cairo was the astronomer 'Abd al-Rahman ibn Yunus (d. 1009). Ibn Yunus was born in Fustat, the predecessor of Cairo, and he witnessed the Fatimid conquest of Egypt as well as the founding of the new capital. He began his astronomical observations in 977, two years after al-Aziz became caliph. When al-Hakim succeeded to the caliphate in 996, at the age of eleven, his keen interest in astrology led him to sponsor Ibn Yunus, who continued his observations until 1003. Ibn Yunus spent the remaining six years of his life completing the *Hakimid Tables,* dedicated to Caliph al-Hakim. These are generally considered to be the most accurate astronomical tables compiled in Islamic science, and their illustrated manuscripts have been compared to those of al-Sufi in their beauty.

Ibn Yunus was also a famous astrologer. The astrological predictions in his treatise *On the Attainment of Desire* are based on the heliacal risings of Sirius (i.e., just before the sun) when the moon is in each of the twelve signs of the zodiac, as well as on the day of the week the Coptic year begins.

A biography of Ibn Yunus by his contemporary al-Musabbihi has been preserved in the works of later writers. This biography reveals that Ibn Yunus was an eccentric who paid no attention to his personal appearance; he was considered a comic figure in Cairo. One day, while in apparent good health, he announced that he would die seven days later, whereupon he locked himself in his house and put his manuscripts in order. He then recited the Koran continuously until he died, passing away on the very day that he had predicted, after which his son sold his manuscripts by the pound in the Cairo soap market.

The most renowned of all the scientists who worked in Fatimid Cairo was Abu 'Ali al-Hasan ibn al-Haytham (ca. 965–ca. 1041), known in the West as Alhazen. Ibn al-Haytham was born in Basra, in Iraq, where he studied mathematics and science before going to Egypt.

Several of his biographers differ about the details of Ibn al-Haytham's life after his departure from Basra. Ibn al-Qifti (d. 1248) says that Ibn al-Haytham went from Iraq to Egypt during the reign of Caliph al-Hakim, to whom he had proposed a construction that would regulate the flow of the Nile. When Ibn al-Haytham arrived in Egypt he was deeply impressed by the many ancient structures along the Nile, and he realized that if a river-control project were at all possible the ancient Egyptians would have put it into effect long ago. He admitted this when he met with al-Hakim, who nevertheless offered him a position in some government office. Ibn al-Haytham accepted the post for fear of offending the caliph, a bloodthirsty tyrant who had executed many of his advisers and commanders. But he sought to avoid dealing with al-Hakim by pretending to be insane, whereupon he was confined to his house and remained there until 1021, when the caliph disappeared one day, riding off into the desert, never to be seen again. Ibn al-Haytham then put aside his pretense of madness and took up residence near the Al-Azhar Mosque, teaching and copying Euclid's *Elements* and Ptolemy's *Almagest*, which supported him while he worked on his researches.

According to another of his biographers, Ibn Abi Usaybi (d. 1227), Ibn al-Haytham in his later years decided to ignore the rest of humanity and devote himself to seeking the truth as the surest way of gaining favor with God, a decision he attributed to his "good fortune, or a divine inspiration, or a kind of madness." His first studies were in theology, but he was so frustrated by this that he became convinced that truth was to be found only in "doctrines whose matter was sensible and whose form was rational." He concluded that such doctrines were to be found in the writings of Aristotle and in works on mathematics, physics, and metaphysics. Ibn Abi Usaybi gives a list of Ibn al-Haytham's works up to 2 October 1038, about three years before his death, consisting of ninety-two titles, including all fifty-five of his extant works. But even that list may not be complete, since Ibn al-Qifti states that he owned a book on geometry in Ibn al-Haytham's own hand dated A.H. 432, or A.D. 1040–1041, probably completed not long before he died.

Ibn al-Haytham's works on logic, ethics, politics, poetry, music, and theology have not survived, nor have his summaries of the writings of Aristotle and Galen. His extant works are in the fields where he is generally agreed to have made his most significant and enduring contributions: astronomy, mathematics, and, particularly, optics.

Ibn al-Haytham's masterpiece is his *Book of Optics*, which is considered one of the most important and influential works ever produced in Islamic science, representing a definite advance beyond what had been achieved by the ancient Greeks in their study of light. The *Optics* was translated into Latin in the late twelfth or early thirteenth century, under the title *Perspectiva*. The *Perspectiva* was the subject of studies and commentaries in Europe until the seventeenth century, stimulating the study of optics in the Latin West.

Book I of the *Optics* presents Ibn al-Haytham's general theory of light and vision. His theory involved "visual rays" projected in straight lines from each point on the surface of a luminous body to a corresponding point on the pupils of the eyes, which act as lenses, from where the optic nerves transmit the "distinct form" of the object to the brain so as to form an image. Book II contains his theory of cognition based on visual perception, which influenced Western philosophers in the fourteenth century. Book III is concerned with binocular vision and with aberrations such as diplopia, or double vision. The next three books deal with catoptrics, phenomena involving reflection, which had been studied by Ptolemy, though not in such exhaustive detail as by Ibn al-Haytham.

The seventh and final book of the *Optics* is devoted to dioptrics, phenomena involving refraction, which also had been studied by Ptolemy. Ibn al-Haytham gives a detailed description of his improved version of Ptolemy's instrument for measuring refraction, which he used to study the bending of light at plane and spherical surfaces with air-water, air-glass, and water-glass interfaces. He summarized the results of his experiments in a set of eight rules for the relation between the angles made by the incident and refracted rays with the normal, or perpendicular to the surface. The last two rules stated that a denser refractive medium bends the light more toward the normal, while a rarer medium bends it away. Ibn al-Haytham was aware, as Ptolemy had been, that these two rules arose from the fact that the velocity of light is greater in the rarer medium than in the denser one. Ibn al-Haytham's theory introduced a new method, that of resolving the velocity of light into two independent components, one along the normal and the other perpendicular to it, where the first component changed in the refraction while the second remained constant. This approach, called the "parallelogram method," was used by a number of European physicists from the thirteenth century onward, in the study of both light and motion.

Ibn al-Haytham refers to the work of an older contemporary named Abu Sa'd al 'Ala ibn Sahl, who is the author of a recently discovered treatise on optics dated 983–85. It is evident from this treatise, and from the reference to him by Ibn al-Haytham, that Ibn Sahl correctly stated the law of refraction, which was not discovered in Europe until the seventeenth century. Although Ibn al-Haytham was aware of Ibn Sahl's discovery, he did not use it in his own study of refraction.

Besides the *Optics*, the extant writings of Ibn al-Haytham include nine minor works on light: *The Light of the Moon, The Halo and the Rainbow, On Spherical Burning Mirrors, The Formation of Shadows, The Light of the Stars, Discourse on Light, The Burning Sphere, The Solar Rays*, and *The Shape of the Eclipse*. The last work is of particular interest because it describes the camera obscura, or pinhole camera, the first appearance of the device that eventually led to the development of photography.

The extant writings of Ibn al-Haytham also include twenty works on astronomy. The most popular of these was the treatise *On the Configuration of the World*, which was translated into Spanish, Hebrew, and Latin. His aim in this work was to give a physical model of the Ptolemaic astronomical system rather than a mathematical theory, one that would be "more truly descriptive of the existing state of affairs and more obvious to the understanding." The model that he chose was that of the homocentric spheres of Eudoxus, which he described fully and clearly without going into unnecessary technical detail, which may be why this work was so popular.

Another of Ibn al-Haytham's extant astronomical writings, known in its Latin translation as *Dubitationes in Ptolemaeum*, is a critique of three of Ptolemy's works: the *Almagest*, the *Planetary Hypotheses*, and the *Optics*. So far as the *Almagest* was concerned, Ibn al-Haytham's main objection was to the equant, which merely disguised the fact that the planets in Ptolemy's model did not move with uniform velocity around the earth as a center

Ibn al-Haytham's fame as a mathematician stems largely from his solution to the so-called Alhazen's problem in Book V of his *Optics*. That is, from two points outside a circle and in its plane, to draw lines meeting at the circumference and making equal angles with the surface normal, or perpendicular, at that point. This leads to a fourth-degree equation, which Ibn al-Haytham solved by finding the intersection points of a circle and a hyperbola.

Aside from the mathematical analysis in the *Optics*, a score of Ibn al-Haytham's writings exclusively on mathematics have survived, most of them brief and varying considerably in importance. One of the longest and most important of these works is entitled *Solution of the Difficulties in Euclid's* Elements. Here he tried to prove Euclid's fifth postulate, defining parallel lines, one of several such attempts by Islamic mathematicians.

Another of his longer mathematical works, *On Analysis and Synthesis*, was written to explain the methods necessary for finding and proving theorems and constructions by illustrating their applications in arithmetic, geometry, astronomy, and music; this work placed particular emphasis on the role of "scientific intuition."

The Fatimid dynasty came to an end in 1171 when Cairo was conquered by the great Kurdish warrior Salah al-Din ibn Ayyub, known in the West as Saladin, founder of the Ayyubid dynasty. Saladin (r. 1171–93) refortified Cairo, building an imposing citadel that still stands, along with the defense walls that enclosed the inner city in his time. Using Egypt as his power base, Saladin went on to conquer Syria and Mesopotamia, defeating the Crusaders at the Battle of Hattin in 1187 and reconquering Jerusalem for Islam.

The leading intellectual figure in Cairo at the beginning of the Ayyubid period was the Jewish philosopher Rabbi Moses ben Maimon, better known in the West as Maimonides (1135–1204). He was born in Cordoba; his family then moved to the Moroccan city of Fez, where he received most of his secular education, studying philosophy, law, and medicine at the Muslim university. In 1166 he settled in Egypt, first in Alexandria and then at Fustat in Cairo, where he became judge and unofficial head of the Jewish community.

After the establishment of the Ayyubid dynasty in 1171, Maimonides became the personal physician to Saladin's vizier, Al-Fadr al-Baisami, and later to Saladin's son and successor, Sultan al 'Aziz (r. 1193–98). At the same time he also tended the sick in Cairo, both Muslims and Jews. Besides his judicial and medical duties, Maimonides spent all of his spare time studying and writing, as he had since his earliest youth.

Maimonides was only fifteen when he finished his first book, *A Treatise on Logic*. This was done in Arabic, as were all of his other books except his *Mishnah Torah,* a codification of the Talmudic law in fourteen volumes written in Hebrew, which took him ten years and was completed in 1180. His earlier books include the astronomical work entitled *Treatise on the Calendar* (1158) and his *Commentary on the Mishnah* (1168). The latter work, besides its discussion of Talmudic law, also contains considerable material on scientific subjects such as zoology, botany, and natural history, as well as psychology.

Maimonides began work in 1185 on what proved to be his masterpiece, *The Guide of the Perplexed,* an explanation of the fundamental theology and philosophy of Judaism, which he finished about five years later. Here Maimonides notes that his purpose is to show that rational philosophy does not contradict Jewish beliefs but, rather, helps a man attain the ultimate state of happiness, which is the perfection of his intellect so that he can contemplate the divine.

His many letters reveal the admiration Maimonides had for both ancient Greek and medieval Islamic philosophers, particularly Aristotle, Plato, al-Farabi, Ibn Sina, and Ibn Bajja, an Andalusian scientist. He accepted Aristotelian physics for the terrestrial world, though not for the celestial realm, which he thought might be beyond human understanding. An even more difficult problem for him was the obvious contradiction between the Aristotelian astronomical model of the homocentric spheres and the mathematical Ptolemaic theory of epicycles, eccentrics, deferents, and equants, and in his own thinking he did not accept any of the attempts by Islamic philosophers and astronomers that sought to resolve these questions.

Two Hebrew translations of *The Guide of the Perplexed* were done shortly after it was written, one by Samuel ibn Tibbon and the other by al-Harizi. During the next three centuries the *Guide* played a central role in Jewish philosophical discussions, with the followers of Maimonides vigorously defending his ideas against his detractors, some of whom wanted his books banned.

The Guide of the Perplexed was translated into Latin in the thirteenth century and exerted a significant influence on the so-called Scholastic philosophy that was developing at that time, as is evident in the works of Saint Thomas Aquinas. The *Guide* was still influential in western Europe as late as the time of Spinoza (1632–1677), who, although he severely criticized Maimonides, agreed with his idea that perfect world peace could be achieved through reason, for this was how Spinoza thought the messianic age would emerge.

Maimonides also wrote extensively on medicine, particularly on diet, psychological treatment, and the use of drugs, and ten of his medical works have survived. He acknowledged his debt to Galen, as did all medieval physicians. Nevertheless, in one of his medical works, *Moses' Chapters on Medicine,* he listed forty contradictions in the writings of Galen, whom he also criticized for being ignorant in philosophy and theology.

Arabic sources rank Maimonides as one of the greatest physicians of all time, particularly because of his skill in treating psychosomatic problems. As an Arabic verse said in his praise, "Galen's medicine is only for the body, but that of [Maimonides] is for both body and soul."

Maimonides was not the only scholar to move across the Islamic world from west to east. The pharmacologist and botanist Ibn al-Baytar (ca. 1190–1248) was born in Malaga and studied in Seville, then moved to Cairo; he died in Damascus. While in Cairo Ibn al-Baytar served as chief herbalist under the Ayyubid sultan al-Kamil (r. 1218–38) and his son and successor al-Salih (r. 1240–49).

Ibn al-Baytar's work in pharmacology is based on the writings of Dioscorides and Galen as well as those of his Arabic predecessors. His two best-known works are *Al-Mughni*, which describes simple medicines used for various illnesses, and *Al-Jami*, an alphabetical list of some fourteen hundred medicines based on his own researches as well as those of his Greek and Arabic predecessors. Ibn al-Baytar's main contribution was his systemization of the researches of Arabic scientists, who added between three and four hundred medicines to the thousand or so known since antiquity. His *Al-Jami* had considerable influence in the East, among both Muslims and Christians, for it was translated from Arabic into Armenian; but it was little known in the West.

The Ayyubid dynasty lasted until 1250, when the last sultan of that line was overthrown by the Mamluks, Turkish slaves who had come to dominate the Egyptian army. Eight years later the Mamluk general Baybars routed the Mongols in a great battle in Syria, the first major defeat suffered by the central Asian nomads, who then retreated into Anatolia and never again directly threatened Egypt. On his return to Egypt Baybars murdered Sultan Qutuz and usurped the throne, beginning one of the longest and most illustrious reigns (1260–76) in the history of the Mamluk dynasty, which lasted until it was annihilated by the Ottoman Turks in 1517.

The court physician during the reign of Sultan Baybars was 'Ala al-Din ibn al-Nafis (ca. 1208–1288), who was born in Transoxiana and studied medicine in Damascus. Besides being a physician, Ibn al-Nafis also lectured on jurisprudence at the al-Masruriyya School in Cairo. His importance as a physician, which led Muslims to call him the "second Ibn Sina," was not fully recognized by Western historians, for many of his medical writings were unknown until quite recent times. His *Compre-*

hensive Book on the Art of Medicine, in eighty volumes, which he wrote in his thirties, was thought to have been lost until 1952, when one fragmentary volume was found in the Cambridge University Library; three other volumes of this work were subsequently discovered in the medical library at Stanford University, one of them dated 1243–44. One of the interesting sections in these fragmentary remains concerns the surgical techniques used by Ibn al-Nafis, which he describes in minute detail, with examples of specific operations as well as discussions concerning the duties of surgeons and the relationships among doctors, nurses, and patients.

The fame of Ibn al-Nafis stems from his discovery of the so-called minor circulation of the blood—that between the heart and lungs. The fact that he had made this discovery was not known until 1924, when the Egyptian physician Dr. Muhyo al-Deen Altawi discovered a manuscript, *The Epitome of the Canon,* an introduction to the work of Ibn Sina in which Ibn al-Nafis first describes the minor circulation of the blood.

It is possible that European physicians first learned of the minor circulation through a translation of the work of Ibn al-Nafis by Andrea Alpago of Belluno (d. 1520). The first European to write about the minor circulation was Michael Servetus (ca. 1510–1553), an Aragonese physician and theologian, who was condemned by Calvin for his unorthodox religious opinions and burned at the stake in Geneva. The definitive theory of blood circulation was finally given by the English physician William Harvey (1578–1657) in his *Exercitatio Anatomica de Motu Cortis et Sanguinis,* published in 1628, which is generally considered to mark the beginning of modern medicine.

Ibn al-Nafis was followed by his student Ibn al-Quff, who won renown as a surgeon and medical writer, his best-known treatise being *The Basic Work Concerning the Art of Surgery.* Ibn al-Quff is credited with being the first to suggest the existence of capillaries in blood circulation. The first European scientist to make this discovery was Marcello Malpighi of Bologna (1628–1694), who in 1661 used a microscope to detect capillaries and explain their role in circulating blood between the arteries and veins.

During the Mamluk period Damascus was the second city of the empire, and in the latter half of the fourteenth century it rivaled and even surpassed Cairo as a center of science, producing one of the most outstanding astronomers in the history of Islamic science, Ibn al-Shatir (ca. 1305–ca. 1375).

Ibn al-Shatir is believed to have been born in Damascus in around 1305. His father died when he was six and he was then brought up by his grandfather, who taught him the craft of inlaying ivory. When he was about ten he traveled to Cairo to study astronomy, in the course of which he was inspired by the work of Abu 'Ali al-Marrakushi, who had, circa 1280, written a compendium of mathematical astronomy and mathematical instruments.

After the completion of his studies Ibn al-Shatir returned to Damascus, where he was appointed chief astronomer of the Umayyad Mosque. His principal duties were to determine the exact times for the five daily occasions of prayer, as well as the dates when the holy month of Ramadan began and ended; he also constructed astronomical instruments and made observations and calculations to compile astronomical tables.

Ibn al-Shatir's first set of tables, which have not survived, apparently used his observations together with the standard Ptolemaic model to compute the entries for the sun, moon, and planets. But in a later work, entitled *The Final Quest Concerning the Rectification of Principles,* he developed an original planetary model significantly different from that of Ptolemy, which he then used to produce a new set of tables in a work called *al-Zij al-Jadid* (The New Planetary Handbook). His preface tells of how he came to write this work after reading the books of earlier Islamic astronomers, who, while expressing doubts about the Ptolemaic planetary model, were unable to formulate an alternative theory.

> I therefore asked Almighty God to give me inspiration and help me to invent models that would achieve what was required, and God—may He be praised and exalted, all praise and gratitude to Him—did enable me to devise universal models for the planetary models in longitude and latitude and all other observable features of their motions, models that were free—thank God—from the doubts surrounding previous models.

Ibn al-Shatir's new planetary model used secondary epicycles rather than the equants, eccentric deferents, and epicycles employed by

Ptolemy, his motive being to have the planets moving in orbits composed of uniform circular motions rather than to improve the agreement of his theory with observation. His model had no advantage over the Ptolemaic theory so far as the sun was concerned, but in the case of the moon it was clearly superior.

There is no evidence that any Arabic astronomer after Ibn al-Shatir formulated a new astronomical theory differing from the Ptolemaic model. His *al-Zij al-Jadid* continued to be referred to in Damascus for several centuries, and it was the subject of commentaries and revisions, one of which adapted it for use in Cairo. The latter was so popular that a commentary on it was published in Cairo in the mid-nineteenth century. Studies by historians of science beginning in 1957 have shown that the lunar model used by Ibn al-Shatir was essentially the same as the one employed by Copernicus in 1543, although research has not revealed the details of how or if in the course of two centuries the new astronomical theory made its way from Damascus to Poland.

8

AL-ANDALUS, MOORISH SPAIN

The Muslim conquest of the Iberian Peninsula began in the spring of 711, when Musa ibn Nusayr, the Arab governor of the Maghreb, or northwest Africa, sent an army across the Strait of Gibraltar under the command of Tariq ibn Ziyad. At that time the Iberian Peninsula was ruled by the Visigoths; their king, Roderick, was defeated and killed in July 711 by Tariq, who went on to capture Cordoba and Toledo, the Visigoth capital.

Musa followed across the strait with an even larger army, and after taking Seville and other cities and fortresses he joined Tariq in Toledo. Musa was then recalled to Damascus by the Umayyad caliph, leaving the conquered lands in the hands of his son 'Abd al-Aziz, who in the three years of his governorship (712–15), extended his control over most of the Iberian Peninsula, which came to known to the Arabs as Al-Andalus.

The first Abbasid caliph, Abu'l-Abbas al-Saffah (r. 749–54), sought to consolidate his power by slaughtering all of the members of the Umayyad family, but one of them, the young prince 'Abd al-Rahman, escaped to the Maghreb and then to Spain, where in 756 he established himself in Cordoba, taking the title of *amir*, or emir. This was the beginning of the Umayyad dynasty in Spain, which was to rule Al-Andalus until 1031. 'Abd al-Rahman I (r. 756–88) established Cordoba as his capital, and in the years 784–86 he erected the Great Mosque, which was rebuilt and enlarged by several of his successors.

The Umayyad dynasty in Al-Andalus reached its peak under 'Abd al-Rahman III (r. 912–61), who in 929 took the title of caliph, emphasizing the independence of Al-Andalus from the Abbasid caliphate in the East. This began the golden age of Muslim Cordoba, known to Arab chroniclers as the "the bride of al-Andalus," referred to by the Saxon nun Hroswith as "the ornament of the world." The golden age continued under 'Abd al-Rahman's son and successor, al-Hakem II (r. 961–76), and his grandson Hisham II (976–1009), who was a puppet in the hands of his vizier al-Mansur, the Latin Almanzor.

'Abd al-Raman chose a site outside Cordoba to build the magnificent palace of Madinat al-Zahra, "the radiant." Al-Hakem built one of the greatest libraries in the Islamic world in Cordoba, rivaling those at Baghdad and Cairo. The caliph's library, together with the twenty-seven free schools he founded in his capital, gave Cordoba a reputation for learning that spread throughout Europe, attracting Christian scholars as well as Muslims, not to mention the Jews who lived under Islamic rule. As the Maghreb historian al-Maqqari was to write of tenth-century Cordoba: "In four things Cordoba surpasses the capitals of the world. Among them are the bridge over the river and the mosque. These are the first two; the third is Madinat al-Zahra; but the greatest of all things is knowledge—and that is the fourth."

After al-Mansur's death in 1002 the caliphate passed in turn to several claimants in the principal cities of Al-Andalus, and finally it was abolished altogether in 1031. The fall of the caliphate was followed by a period of sixty years in which Al-Andalus was fragmented into a mosaic of petty Muslim states, allowing the Christian kingdoms of northern Spain to start expanding south, beginning what came to be known as the Reconquista. The first major Christian triumph came in 1085, when Toledo fell to the king of Castile and León, Alfonso VI (r. 1072–1109).

The fall of Toledo led the petty Muslim rulers to seek help from the powerful ruler of the Almoravids in Morocco, Yusuf ibn Tashfin (r. 1061–1106). Yusuf crossed into Al-Andalus in 1086 and decisively defeated Alfonso's army, saving southern Spain from falling into Christian hands. This led to the domination of Al-Andalus by the Almoravids, which lasted until the mid-twelfth century, when they were supplanted by another powerful dynasty from the Maghreb, the Almohads. During the reign of 'Abd al-Mu'min (r. 1130–63) the Almohads extended their power throughout both the Maghreb and Al-Andalus. The Almohads suffered a crushing defeat in 1212 at the hands of a Christian coalition, which in the next half century seized the major Muslim cities in Al-Andalus, taking Cordoba in 1236. Virtually all that remained of Muslim Spain was the Banu Nasr kingdom of Granada, which hung on until its capture in 1492 by Ferdinand II of Aragon and Isabella of Castile, the "Catholic Kings," who drove the last of the Moors from Spain.

'Abd al-Rahman II (r. 822–52) began the development of science in Al-Andalus by sending an agent to the East to buy books, which an anonymous Maghreb chronicler says included astronomical tables as well as works in astronomy, philosophy, medicine, and music. The *amir* was keenly interested in astronomy and astrology, perhaps stimulated by a total eclipse of the sun on 17 September 833, which so terrified the people of Cordoba that they quickly gathered at the Great Mosque to pray for divine deliverance.

The *amir*'s court poet and astrologer was 'Abbas ibn Firnas (d. 887), who introduced a version of al-Khwarizmi's astronomical tables, the *Zij al-Sindhind*. With the *amir*'s patronage, Ibn Firnas built an observatory in Cordoba, with a planetarium, an armillary sphere, and a water clock capable of indicating the times of prayer. He also attempted to fly by leaping from the top of the Rusafa palace in Cordoba with a hang glider of his own invention. He apparently managed to glide for some distance but suffered injuries in a rough landing, which his critics attributed to his failure to observe the manner in which birds use their feathers when they alight on a branch.

Tenth-century Cordoba was renowned for its school of physicians, presided over by the Jewish doctor Hasday ibn Shaprut, vizier of 'Abd al-Rahman III and later personal physician of Hisham II. Hasday also supervised the imperial translation activities and carried out diplomatic missions on behalf of the caliphate. One of his diplomatic activities involved the reception of an ambassador from the Byzantine capital Constantinople in 949. The envoy brought with him presents for 'Abd al-Rahman III from the emperor Constantine VII Porphyrogenitus (r. 913–59), one of them being a superb Greek manuscript of Dioscorides' *De Materia Medica*.

No one in Cordoba knew enough Greek to read the manuscript, so the ambassador arranged for a Byzantine monk named Nicholas to be sent to Cordoba. He arrived in 951, along with a Greek-speaking Arab from Sicily. Nicholas and the Arab then explained Dioscorides' work to a group of Cordoban scholars headed by Hasday, thus beginning the study of pharmacology in Al-Andalus. *De Materia Medica* was subsequently translated from Arabic into Latin for the education of pharmacists and physicians in Christian Europe.

The principal source of information about the Cordoban medical school is Ibn Juljul al-Andalusi (944–ca. 994), who studied medicine there between the ages of fourteen and twenty-four. His most important work, entitled *Generations of Physicians and Wise Men,* is the most complete extant source in Arabic on the history of medicine. He says that most of the physicians practicing in Al-Andalus up to the time of 'Abd al-Rahman III were Mozarabs, or Christians living under Arab rule, and that the principal source of their knowledge was "one of the books of the Christians that had been translated," which would have been Dioscorides' work.

Ibn Juljul also wrote a treatise on Dioscorides' *De Materia Medica,* probably based on the manuscript that had been sent from Constantinople. The works of Ibn Juljul remained popular in Al-Andalus for centuries, and one of them may have been translated into Latin, since Albertus Magnus quotes from a treatise that he attributes to a certain Gilgil, probably a corruption of "Juljul."

The physician Abu'l Qasim al-Zahrawi (ca. 936–ca. 1013), the Latin Abulcasis, was a contemporary of Ibn Juljul's. He lived in the imperial Cordoban suburb of Madinat al-Zahra, his family tracing its origins back to the Arab warriors who originally conquered Al-Andalus. His only known work is the *Kitab al-Tasrif,* a medical encyclopedia in thirty volumes, which he completed in about the year 1000, encompassing nearly half a century of experience as a physician. The encyclopedia covers every aspect of medicine, including the design and manufacture of surgical tools, midwifery, pharmaceutical preparations, diet, hygiene, medical terminology, weights and measures, medical chemistry, human anatomy and physiology, therapeutics, and psychotherapy. Al-Zahrawi particularly emphasized the importance of bedside manner and the bond between doctor and patient, writing, "Only by repeated visits to the patient's bedside can the physician follow the progress of his medical treatment."

Al-Zahrawi was also a pioneer in the use of drugs in psychotherapy, and he made an opium-based medicine that he called "the bringer of joy and gladness, because it relaxes the soul, dispels bad thoughts and worries, moderates temperaments, and is useful against melancholia."

The development of astronomy in Al-Andalus begins with the work of Abu Maslama al-Majriti, who was born in Madrid and studied in Cordoba, where he died in 1007. Together with his student Ibn al-Saffar (d. 1034), he improved the astronomical tables of al-Khwarizmi and adapted them for the longitude of Cordoba, a work that passed to Christian Europe through a Latin translation by Adelard of Bath. Two other extant works of al-Majriti's are the *Commercial Arithmetic* and the brief *Treatise on the Astrolabe*, while his Arabic translation of Ptolemy's *Planisphaerium* survives in a Latin version by Herman of Dalmatia. The eleventh-century historian Ibn Sa'id of Toledo says that al-Majriti "applied himself to the observation of the heavenly bodies and to understanding the book of Ptolemy called the *Almagest*," and that he was "the author of a summary of the part of al-Battani's table concerning the equation of the planets."

Another work attributed to al-Majriti is the *Ghayat al-Hakim* (The Aim of the Wise), which was translated into Spanish in 1256 through the patronage of King Alfonso X of Castile. It was later translated into Latin under the title *Picatrix*, a corruption of Buqratis, the Arab name of Hippocrates, on the supposition that he, and not al-Majriti, was the author, who is described on the title page as being a "very wise ... philosopher ... most skilled in mathematics ... [and] very learned in the arts of necromancy."

The *Picatrix* has been described as "a compendium of magic, cosmology, astrological practice, and esoteric wisdom in general" that "provides the most complete picture of superstitions current in eleventh-century Islam." Lynn Thorndike devotes a whole chapter of his *History of Magic and Experimental Science* to the *Picatrix*, which he describes as a "confused compilation of extracts from occult writings and a hodgepodge of innumerable magical and astrological recipes."

Al-Majriti had a number of students who spread his knowledge of both science and magic throughout Al-Andalus and even beyond, the best known being Ibn al-Samh of Granada (d. 1035), al-Kirmani of Saragossa (d. 1055), and Ibn al-Saffar. Knowledge of al-Majriti's work evidently made its way to the eastern Islamic world, for nearly four centuries later he is mentioned by Ibn al-Shatir of Damascus as one of those who had produced astronomical models that were different from the standard Ptolemaic theories.

The leading Andalusian astronomer in the century after al-Majriti was Ibn Mu'adh al-Jayyani (d. 1093), whose last name comes from the fact that he was a native of Jaén, east of Cordoba. His best-known work is the *Tabulae Jahenen*, a set of astronomical tables based on al-Khwarizmi's *Zij al-Sindhind* and adapted for the longitude of Jaén. His tables were an improvement over the *Sindhind*, for he took into account the precession of the equinoxes, which al-Khwarizmi had ignored, and he utilized advances in astronomical theory made by al-Biruni and his other predecessors. The *Tabulae Jahenen* also gives detailed instructions in such practical matters as determining the times of prayer, the direction of Mecca, the beginning of the Islamic months, and the casting of horoscopes, all of which made it very helpful for mosque astronomers.

Al-Jayyani's other writings include treatises on astronomy and mathematics. His astronomical works include a treatise dealing with the phenomena of twilight and false dawn, which in its Latin translation was popular from the medieval era until the Renaissance. One of his mathematical works is a treatise on spherical trigonometry. Another is his treatise *On Ratio*, which he says he composed "to explain what may not be clear in the fifth book of Euclid's writing to such as are not satisfied with it." But there is no evidence that he did clarify the fifth book of the *Elements*.

Al-Jayyani's treatise on spherical trigonometry was indirectly transmitted to Christian Europe through a work of Jabir ibn Aflah's, an astronomer and mathematician who flourished in Seville in the first half of the twelfth century. Ibn Aflah's most important work, in which he used and added to al-Jayyani's methods in spherical trigonometry, is an adaptation of Ptolemy's astronomical theories in a treatise entitled *Islah al-Majisti* (Correction of the *Almagest*). The *Islah* was translated into Latin and Hebrew, and it was used by Muslim, Jewish, and Christian astronomers and mathematicians until the seventeenth century.

Another set of astronomical tables was compiled for Toledo around 1069. These were the famous *Toledan Tables,* known only through a Latin translation, which survives in an enormous number of manuscript copies. The tables, which were an adaptation of earlier works from Ptolemy through al-Khwarizmi and al-Battani, were prepared by a group of astronomers, the best-known of whom was Abu'l-Qasim Sa'id (d. 1070), the *qadi*, or judge, of Toledo.

Another notable member of the group was Ibn al-Zarqali (d. 1100), the Latin Arczachel, a self-educated artisan who worked for Abu'l-Qasim Sa'id as a maker of astronomical instruments and water clocks. After Abu'l-Qasim Sa'id died, al-Zarqali became director of the group that completed the new astronomical tables. The *Toledan Tables* were used in both Al-Andalus and in Christian Europe, where they were translated into Latin circa 1140 as the *Marseilles Tables*. They remained in use until the fourteenth century, and a Latin version of the *Toledan Tables* was translated into Greek, completing a remarkable cultural cycle. The tables are mentioned by Chaucer in "The Franklin's Tale," where one of the characters is a magician-astrologer of Orleans, equipped with all the tools of his celestial trade:

> His tables Toletanes forth he brought
> Ful wel corrected, ne ther lacked noght,
> Neither his collect ne his expans yeres,
> Ne his rotes ne his othere geres.

The observations that led to the *Toledan Tables* were continued for another three decades by al-Zarqali, who in around 1078 left Toledo because of the repeated attacks by the Christian king Alfonso VI and moved to Cordoba, where he lived for the rest of his days. The water clocks built in Toledo by al-Zarqali remained in use until 1133, when King Alfonso VII of Castile and León had them taken apart to see how they worked but could not reassemble them. Water clocks of the type built by al-Zarqali, which showed the motion of the celestial bodies, became popular in seventeenth-century Europe.

Other works by al-Zarqali include six treatises on mathematical astronomy and astronomical instruments. One of these treatises is an adaptation of an astronomical work by Ammonius of Alexandria entitled the *Almanac*. Al-Zarqali's elaboration of the *Almanac*, in which he uses Babylonian astronomy as well as drawings from the works of Hipparchus and Ptolemy, was translated into Latin, Hebrew, Portuguese,

Catalan, and Castilian, and it remained in use until the fifteenth century. Another of his astronomical treatises describes the orbit of Mercury as "oval" rather than circular, which a modern Islamic scholar has interpreted as anticipating Kepler's theory of elliptical orbits, but this seems highly unlikely.

The beginning of Arabic philosophy in Al-Andalus comes with the work of Ibn Hazm (994–1064), who was born and spent most of his life in Cordoba, where his father and grandfather had been functionaries in the Umayyad court. His best-known philosophical work is his *Book on the Classification of the Sciences*. Aside from his many philosophical compositions, he also wrote poetry and treatises on history, jurisprudence, ethics, and theology. His most famous poetical work is entitled *Tawq al-Hamama* (The Dove's Neck-Ring), a treatise on the art of love, which he says is "a serious illness."

> *I've a sickness doctors can't cure,*
> *Inexorably pulling me to the well of my destruction.*
> *Consented to be a sacrifice, killed for her love,*
> *Eager, like the drunk gulping wine mixed with poison.*
> *Shameless were those my nights,*
> *Yet my soul loved them beyond all passion.*

Ibn Hazm notes that he was particularly qualified to write a book on the art of love, having been brought up until the age of fourteen in the harem, or women's quarters, of his family home: "I have observed women at first hand and I am acquainted with their secrets to an extent that no one else could claim, for I was raised in their chambers and I grew up among them and knew no one but them." He goes on to say that "women taught me the Koran, they recited to me much poetry, they trained me in calligraphy."

The Islamic schools of the time in Cordoba employed scores of women copyists, as did the city's book market. More highly educated women worked as teachers and librarians, and a few even practiced medicine and law.

The next Andalusian philosopher of note after Ibn Hazm was Ibn Bajja, known in Latin as Avempace. Ibn Bajja was born in Saragossa circa 1070, and in the years 1110–18 he served as vizier to the Almoravid governor of the city, Ibn Tifilwit. After the Christian conquest of Saragossa he spent the rest of his life in Almoravid territory, living in turn in Almeria, Granada, and Seville. While in Seville he was imprisoned before being released due to the intervention of Ibn Rushd al-Jadd, grandfather of the philosopher Averroës. After his release he moved first to Jaén and then to Fez in Morocco, where he died in 1128. Tradition says that he was killed by eating an eggplant poisoned by his rivals, intellectuals in the Almoravid court in Fez.

Thirty-seven of Ibn Bajja's numerous works survive, many of them commentaries on the works of Aristotle, Euclid, Galen, and al-Farabi, along with three of his own works. His ideas influenced the thought of Ibn Tufayl (Abubacer), Ibn Rushd (Averroës), Maimonides, and, in Latin translation, Saint Thomas Aquinas.

Ibn Bajja seems to have been the first Arabic scientist in Al-Andalus to oppose the Ptolemaic planetary model. He rejected the use of epicycles as being incompatible with Aristotle's doctrine of celestial motion, in which the planets move in perfect circles about the earth, at the center.

Ibn Bajja's ideas on dynamics appear in his notes on Aristotle's *Physics*. Here he rejected the Aristotelian law of motion, which held that the velocity of a body was directly proportional to the motive power and inversely proportional to the resistance of the medium through which it moved. Instead, following John Philoponus, he said that motion would occur only when the motive power was greater than the resistance, and that the velocity was proportional to the difference between the power and the resistance. This meant that in a void a body would move with finite speed, rather than infinitely fast. He argued further that even in a void a body had to traverse a definite distance in any given time, so that its velocity would be finite no matter how fast it was moving. This was counter to the Aristotelian notion that in a vacuum a body's velocity would be infinite, which was impossible, so that a void could not possibly exist.

Ibn Bajja was also an accomplished musician and poet. According to the thirteenth-century Tunisian writer al-Tifashi, Ibn Bajja "combined the songs of the Christians with those of the East, thereby inventing a style found only in Andalus, toward which the temperament of its people inclined so that they rejected all others."

Abu Marwan ibn Zuhr (ca. 1092–1162), the Latin Avenzoar, was the most famous of a family of Seville physicians who served the Almoravid dynasty in Al-Andalus and the Maghreb. Ibn Zuhr served as personal physician to the *amir* 'Ali ibn Tashfin (r. 1106–43) in his palace at Marrakesh, but because of a misunderstanding he was imprisoned by his patron. When the Almoravids were overthrown by the Almohads, Ibn Zuhr was restored to favor by the new ruler, Abd al-Mu'min (r. 1145–63), who appointed him as his court physician and personal counselor, with the rank of emir.

Ibn Zuhr's medical writings were based on the works of Hippocrates and Galen as well as those of his Arabic predecessors and his own researches. His best-known work, *An Aid to Therapy and Regimen*, was translated into Hebrew and Latin and remained in use until the European Renaissance. Ibn Zuhr was generally considered to be one of the best physicians in Al-Andalus, particularly as a clinician and medical therapist.

Abu Bakr Muhammed ibn Tufayl (c. 1110–85), a student of Ibn Bajja's, was the personal physician and vizier to the Almohad caliph Abu Ya'qub Yusuf (r. 1163–84), the builder of the Great Mosque in Seville, his capital. Ibn Tufayl continued the tradition of his teacher Ibn Bajja in opposing the Ptolemaic planetary theory. He apparently formulated a planetary model that avoided using the eccentrics and epicycles of Ptolemy. Ibn Tufayl was the first Andalusian thinker to make use of the works of Ibn Sina, though with some differences, such as his belief that there is no proof that the world is eternal rather than created in time.

The culmination of Arabic philosophy comes with Ibn Rushd, the Latin Averroës (1126–98), who was from a distinguished family of Cordoban jurists. He was named for his grandfather, who was imam of the Great Mosque and also *qadi*, a position his father also held. He studied theology, law, medicine, and philosophy, including the works of Aristotle, particularly his writings in physics and natural science.

Ibn Rushd was in Marrakesh in 1152, during the reign of the Almohad ruler 'Abd al-Mu'min, when he seems to have made his first astronomical observations. There he may have met Ibn Tufayl, who would later play an important part in his life by introducing him to the caliph Abu Ya'qub Yusuf. According to Bundud ibn Yahya, a disciple of Ibn Rushd's, the caliph had complained to Ibn Tufayl about his difficulty in reading the works of Aristotle and the need for a commentary to explain them. Ibn Tufayl said that he himself was too old and busy to do the job, and so he recommended Ibn Rushd, who was thus led to begin his monumental commentary on the works of Aristotle.

After the death of Ibn Tufayl, Ibn Rushd became the personal physician to Abu Ya'qub Yusuf and was appointed *qadi,* first in Seville, then in Cordoba, and then again in Seville. He retained his posts under Abu Ya'qub Yusuf's son and successor Abu Yusuf Ya'qub al-Mansur (r. 1184–99), though in 1195 the caliph confined him for two years to the town of Lucena, near Cordoba, because fundamentalist Islamic scholars had condemned his philosophical doctrines. Early in 1198 the caliph lifted the ban and took Ibn Rushd with him to his court at Marrakesh. But Ibn Rushd had little time to enjoy his freedom, for he died in Marrakesh on 10 December of that year, after which his body was returned to Cordoba for burial.

The philosophical writings of Ibn Rushd can be divided into two groups: his commentaries on Aristotle and his own treatises on philosophy, entitled *Decisive Doctrine About the Concordance Between Revelation of Religion, Exposition of the Methods of Demonstration,* and *Incoherence of the Incoherence of the Philosophers.* The last treatise was written in opposition to al-Ghazali's attack on rational philosophy, particularly the works of al-Farabi and Ibn Sina, the two leading Muslim interpreters of Aristotle. Here, in his defense of Aristotelianism, Ibn Rushd shows how al-Farabi and Ibn Sina often deviated from the ideas of Aristotle, trying to resolve the dispute between Islamic theologians and philosophers and to reconcile apparent contradictions between Scripture and science. His commentaries attempted to restore Aristotle's own ideas in Islamic thought and to supplant the Neoplatonism of al-Farabi and Ibn Sina. He regarded the philosophy of Aristotle as the last word, to the extent that truth can be understood by the human mind.

Ibn Rushd's writings deeply influenced Maimonides and, through him, other Jewish scholars, who read his works in Arabic. By the beginning of the thirteenth century Ibn Rushd was considered to be the outstanding interpreter of Aristotle and his works were translated into Hebrew. By the end of that century nearly half of his commentaries on Aristotle had been translated from Arabic into Latin, so that he came to be known in the West as the "Commentator."

Following earlier Arabic philosophers, Ibn Rushd interpreted the concept of creation in such a way as to deny free will not only to man but even to God himself. According to Ibn Rushd, the world had been created by a hierarchy of necessary causes, starting with God and descending through the various "Intelligences" that moved the celestial spheres. Each of the eight spheres—those of the stars, sun, moon, and five planets—has its own incorporeal intelligence, which serves as what Ibn Rushd calls its "own object of desire," since each performs its own unique motion.

Ibn Rushd accepted Aristotle's planetary model of the homocentric spheres and rejected Ptolemy's theory of eccentrics and epicycles. He writes of his astronomical researches in his commentary on Aristotle's *Metaphysics*, where he expresses his belief that the prevailing Ptolemaic theory is a mathematical fiction that has no basis in reality.

In his commentary on Aristotle's *Physics*, Ibn Rushd attacked Ibn Bajja's theory of motion, specifically the idea that the medium impeded natural motion. Instead he supported Aristotle's theory, in which the velocity of a body is proportional to the force acting on it divided by the resisting force of the medium. Actually, both theories are incorrect; the first correct explanation came with Newton's laws of motion in 1687.

Ibn Rushd's major work on medicine is his *Al-Kulliyyat* (Generalities), which is based on the writings of Galen. He was a very close friend of Ibn Zuhr, who dedicated his *Al-Taisir* to him. Ibn Rushd's *Al-Kulliyyat* and Ibn Zuhr's *Al-Taisir* were meant to constitute a comprehensive medical textbook, and some Latin editions contain both treatises bound together as a single book, which in some places supplanted Ibn Sina's *Canon*. One of the discoveries made by Ibn Rushd in his medical researches was that the retina rather than the lens is the sensitive element in the eye, an idea that was forgotten until it was revived by the anatomist Felix Platter (1536–1614).

Ibn Rushd was the first writer in any language to complain about discrimination against women, which he felt was one of the most serious problems in Muslim society.

Our society allows no scope for the development of women's talents. They seem to be destined exclusively to childbirth and the care of children, and this state of servility has destroyed their capacity for larger matters. It is thus that we see no women endowed with moral virtues, they live their lives like vegetables, devoting themselves to their husbands. From this stems the misery that pervades our cities, for women outnumber men by more than double and cannot procure the necessities of life by their own labors.

Ibn Tufayl's researches in astronomy were continued by his student al-Bitruji (fl. ca. 1190), the Latin Alpetragius, whose only known work is his *Kitab fi'l-Hay'a* (Book of Astronomy). Al-Bitruji acknowledged that Ptolemy's theory gave an exact mathematical description of planetary motion. But he felt that the Ptolemaic model was unsatisfactory since its eccentrics, epicycles, and equants were incompatible with Aristotle's physical concept of the homocentric spheres. And so he tried to formulate a model in which a simple system of concentric spheres, one for each planet, would give results equivalent to those of Ptolemy's theory.

The *Kitab fi'l-Hay'a* was translated into Hebrew and Latin, leading to the spread of al-Bitruji's ideas through much of Europe from the thirteenth century into the seventeenth. Al-Bitruji's planetary model was used by those who defended Aristotle's theory of the homocentric spheres against the supporters of Ptolemy's eccentric epicycles and equants. Isaac Israeli (fl. 1310) of Toledo seems to be referring to al-Bitruji when he writes of "the man whose theory shook the world." Copernicus refers to al-Bitruji in connection with the order of the planets Mercury and Venus in his heliocentric theory of 1543.

After the fall of Cordoba to the Christians in 1252, western Arabic science continued in Granada, the last Muslim kingdom in Al-Andalus, and in the Maghreb, though on a much diminished scale.

The mathematician Ibn al-Banna al-Marrakushi (1256–1321) was a native of Granada, though, as his last name indicates, he seems to have spent most of his life in Marrakesh. He is known to have studied in both Marrakesh and Fez, where he taught mathematics and astronomy in the madrasa al-Attarin. Eighty-two of his works are known, of which the most important is the *Summary of Arithmetical Operations*, a compendium of the lost works of the mathematician al-Hassar (fl. ca. 1200).

Al-Qalasadi (ca. 1412–ca. 1506) is the last Arabic mathematician known to have lived in Al-Andalus and the Maghreb. He was a native of

Basta (now Baza) in Spain, but when the city was taken in 1486 by Queen Isabella of Castile he was forced to flee to the Maghreb, where he died at Beja in Tunisia. One of al-Qalasadi's works is a commentary on Ibn al-Banna's *Summary of Arithmetical Operations*. The first of his own writings was the *Classification of the Science of Arithmetic*, which he followed with a simplified version entitled *Unveiling the Science of Arithmetic*, and then an abridgment of this work called *Unfolding the Secrets of the Use of Dust Letters* (i.e., Hindu numerals). The last two works were used in Moroccan schools for generations after the death of al-Qalasadi.

Al-Qalasadi died only a few years after the fall of Granada in 1492, which ended the history of Al-Andalus. The principal remnant of the intellectual world of Muslim Granada is the Casa de la Ciencia (House of Science), founded in 1349 by the *amir* Yusuf I (r. 1334–54). Only fragments of the Moorish building remain, but it is still referred to by its original Spanish name, La Madraza, from *madrasa*, the Arabic word for a Muslim school of higher studies, the last one in Al-Andalus. La Madraza was the predecessor of the University of Granada, founded in 1531 by the emperor Carlos V, as Christian Spain picked up the study of science from where it had been advanced by Muslim Al-Andalus.

FROM TOLEDO TO PALERMO:
ARABIC INTO LATIN

I slamic science in Al-Andalus was still at its peak when the first
Christian scholars came to study in Spain. There they learned science from Arabic sources and translated it into Latin, often in collaboration with local multilingual scribes, mostly Jewish, some of
whom voluntarily converted to Christianity. At the same time other
scholars were at work across Europe, from Toledo to Palermo, translating from Arabic into Latin as well as writing original scientific treatises.

The earliest evidence of European acquisition of Islamic science is a
tenth-century Latin manuscript from the library of the monastery of
Santa María de Ripoll in Catalonia, now in the archives of the Crown of
Aragon in Barcelona. The manuscript begins with a brief treatise on the
astrolabe and contains a table of the brightest stars, which are referred
to by the Arabic names by which they are still known today, such as
Altair, Vega, Rigel, Aldebaran, and Algol. Another tenth-century Latin
manuscript, preserved in the Bibliothèque Nationale in Paris, is entitled
Mathematica Alhandrei Summi Astrologi (Mathematics of Alhandreus,
Supreme Astrologer). "Alhandreus" appears to be a corruption of
"Alkindes," the Latin for al-Kindi, the ninth-century Islamic philosopher. He writes in the preface, "These are the twenty-eight principal
parts or stars [i.e., constellations] through which the fates of all are dis-
posed and pronounced indubitably, future as well as present. Anyone
may with diligence forecast goings and returning, origins and endings,
by the most agreeable aid of these horoscopes."

The first major figure in the European acquisition of Greco-Arabic science is Gerbert d'Aurillac (ca. 945–1003), who became Pope Sylvester II (r. 999–1003). Gerbert's writings include a letter he sent in May 984 to a certain Lupitus of Barcelona, whom he asked to send a translation he had made of a treatise on astrology, presumably from an Arabic work. Gerbert himself is credited with a treatise on the astrolabe entitled *De Astrolabia*, as well as the first part of a work entitled *De Utilitatibus Astrolabi*, both of which show Arabic influence. His attested writings also include works on mathematics, one of which is a treatise on the abacus, a calculating instrument that is believed to have come to the Islamic world from China, where it is still in use. He also constructed a device to represent the celestial sphere, which he used in his classes on astronomy in the cathedral school at Rheims. Gerbert's students are known to have gone on to teach at eight other cathedral schools in northern Europe, where his inspiration led them to emphasize the mathematical sciences he had learned from Islamic sources in Spain.

Gerbert later acquired the reputation of being a magician, a legend that seems to have begun in the first half of the twelfth century with William of Malmesbury. William says that Gerbert fled from his monastery to study astrology and the black arts with the Saracens, from whom "he learned what the song and flight of birds portend, to summon ghostly figures from the lower world, and whatever human curiosity has encompassed whether harmful or salutary." A thirteenth-century manuscript in the Bodleian Library at Oxford says that Gerbert became archbishop and pope with the aid of demons, and that he had a genie in a golden head whom he consulted when solving difficult mathematical problems.

Another key figure in the early transmission of Arabic science to the Latin West is Hermannus the Lame (1013–1054), a son of Count Wolferat of Althausen, in southern Germany. Hermannus is one of the earliest Latin authors to introduce to the Latin West three astronomical instruments that had been widely used in the Islamic world: the astrolabe, the chilinder, and the quadrant. These are described in *De Mensure Astrolabi* and *De Utilitatibus Astrolabi*, two works that have been attributed to Hermannus, though the first part of the latter work may be by Gerbert d'Aurillac. All three instruments became widely used in the Latin West for astronomical observations as well as for calculations.

Other works by Hermannus include a primer to teach multiplication and division with the abacus, using only Roman numbers. He also wrote the earliest-known treatise on *rithmomachia*, a complex board game based on Pythagorean numbers that was very popular in the Latin West during the Middle Ages.

The first of the important translators of Greco-Islamic science from Arabic into Latin is Constantine the African (ca. 1020–1085). An account of his early life is given by a twelfth-century Salerno physician known only as Magister Mattheus F. According to this account, Constantine was a Muslim merchant from Carthage, in North Africa, who visited the Lombard court at Salerno in southern Italy, where he learned that there was no medical literature available in Latin. He went back to North Africa and studied medicine for three years, after which he returned to Salerno, perhaps as early as 1065, with a collection of medical writings in Arabic. A few years later he converted to Christianity and became a monk in the Benedictine abbey at Monte Cassino. There, under the patronage of the famous abbot Desiderius, later Pope Victor III, he spent the rest of his days making Latin translations and compilations from Arabic medical texts.

Petrus Diaconus, the historian of the monastery at Monte Cassino, lists a score of translations by Constantine, including works by Hippocrates and Galen as well as those of the Jewish physician Isaac Israeli and the Arabic writers Ibn al-Jazaar and al-Majusi. His most ambitious work was al-Majusi's *Kitab al-Maliki*, which he translated as the *Pantegne*, divided into two ten-chapter sections, "Theorica" and "Practica," suppressing the name of the author and thus leaving himself open to charges of plagiarism. Constantine appears to have translated only about half of this work, which seems to have been completed by his student Johannes Afflacius.

There is no direct evidence to connect Constantine with the Medical School of Salerno, founded in the mid-eleventh century. Johannes Afflacius seems to have taught there and introduced Constantine's translations into the curriculum under the title of *Ars Medicine* or *Articella*, which formed the foundation of a large part of European medical education on into the sixteenth century. Constantine had always emphasized that medicine should be taught as a basic part of natural philosophy, and the "Theorica" section of the *Pantegne* provided the basis for this integrated study.

The Jewish scholar Isaac Israeli flourished in Tunisia in the first half of the tenth century, where he was court physician to the last Aghlabid emir and then to the Fatimid caliph who succeeded him. His three most important medical works are the *Book on Fevers*, the *Book on Urine*, and the *Book on Foodstuffs and Drugs*, all three of which were translated into Latin by Constantine the African. The first two of these works were popular as textbooks and were also translated into Hebrew. Isaac also wrote a number of short works on philosophy. The best-known of these was the *Book of Definitions and Descriptions*, largely based on the work of al-Kindi; it was translated into Latin by Gerard of Cremona and became a popular textbook in the first European universities. The book is a collection of fifty-seven definitions, most of them paraphrases and quotes from al-Kindi's terminology, which is used for both terrestrial and celestial objects.

The First Crusade, which began in 1096, led to the establishment of Crusader states in Edessa, Antioch, and Jerusalem, an important factor in opening up Islamic culture to western Europe. One of the earliest examples of this cross-cultural contact is the work of Stephen of Antioch, a translator who flourished in the first half of the twelfth century. According to Matthew of Ferrara, Stephen was a Pisan who went to Syria, probably to the Pisan quarter of Antioch, where his uncle was the Roman Catholic patriarch.

At Antioch Stephen learned Arabic and translated the medical treatise *Kitab al-Maliki* of al-Majusi into Latin, under the title of *Regalis Dispositio*; he completed it in 1127. Stephen said that he did so because he felt that the previous translation of this work, by Constantine the African, was incomplete and distorted. He also added a prologue to the second part of the treatise, a list of synonyms in three columns—Arabic, Latin, and Greek—as an aid to help his readers understand the Arabic terms in Dioscorides' *De Materia Medica*. There he noted that those who have difficulty with the Latin terms can consult experts, "for in Sicily and Salerno, where students of such matters are chiefly to be found, there are both Greeks and men familiar with Arabic."

Stephen wrote that the *Regalis Dispositio* was his first work, and he went on to say that he hoped to translate a portion of "all the secrets of philosophy that lie hidden in the Arabic tongue." This has led to the suggestion that he may be the Stephen Philosophus who wrote several books on astronomy based on Arabic and Greek sources.

Adelard of Bath (ca. 1080–1152) was one of the leading figures in the

European acquisition of Arabic science. In the introduction to his *Questiones Naturales*, addressed to his nephew, Adelard writes of his "long period of study abroad," first in France, where he studied at Tours and taught at Laon. He then went on to Salerno, Sicily, Asia Minor, Syria, and, probably, Palestine and Spain. It was most likely in Spain that Adelard learned Arabic, for his translation of al-Khwarizmi's *Sindhind* was from the version revised by the Andalusian astronomer Maslama al-Majriti. Adelard's translation, comprising thirty-seven introductory chapters and 116 listings of celestial data, provided Christian Europe with its first knowledge of Greco-Arabic-Indian astronomy and mathematics, including the first tables of the trigonometric sine function to appear in Latin.

Adelard was also the first to translate the full text of Euclid's *Elements* into Latin, beginning the process that led to Euclid's domination of medieval European mathematics. He did three versions of the *Elements*, the first being from the Arabic of al-Hajjaj, who had translated it from Greek for Caliph Harun al-Rashid. The second was an abbreviated version that Adelard called a *commentum*, in which, among other elaborations, he gave "enunciations" (explanations) of the definitions, postulates, axioms, and propositions in Book I of the *Elements*. The third version, which Roger Bacon referred to as an *editio specialis*, adds more *commenta*, along with full proofs of all propositions.

Companus of Novara adapted Adelard's second version to produce what is considered to be the best Arabic to Latin translation of the *Elements*, the earliest extant copy of which is dated 1259. This and the first translation of the *Elements* from Greek to Latin, by Bartolomeo Zamberti in 1505, formed the basis for most subsequent versions, including those in the vernacular languages of Europe. The first introduction to the *Elements* in English appears in Robert Recorde's *The Pathway to Knowledge*, published in London in 1551. Recorde realized that Euclid's axioms would be far beyond the mathematical ability of the "simple ignorant" people who would read his book, "For nother is there anie matter more straunge in the English tunge, than this whereof never booke was written before now, in that tungue." The first complete English translation, published in London in 1570, was by Sir Henry Billingsley, later lord mayor of London, with a "fruitfull Praeface" by John Dee, who writes that the book contains "manifolde additions, Scholies, Annotations and

Inventions . . . gathered out of the most famous and chiefe Mathematicians, both of old time and in our age."

Adelard says that his *Questiones Naturales* was written to explain "something new from my Arab studies." The *Questiones* are *seventy-six* in number, 1–6 dealing with plants, 7–14 with birds, 15–16 with mankind in general, 17–32 with psychology, 33–47 with the human body, and 48–76 with meteorology and astronomy. Throughout he looks for natural rather than supernatural causes of phenomena, a practice that would be followed by later European writers. His observations are for the most part accurate, and he remarks that he prefers reason to authority.

One particularly interesting passage in this work comes when Adelard's nephew asks him if it were not "better to attribute all the operations of the universe to God." Adelard replies: "I do not detract from God. Everything that is, is from him and because of him. But [nature] is not confused and without system and so far as human knowledge has progressed it should be given a hearing. Only when it fails utterly should there be a recourse to God."

The *Questiones Naturales* remained popular throughout the rest of the Middle Ages, with three editions appearing before 1500, as well as a Hebrew version. Adelard also wrote works ranging from trigonometry to astrology and from Platonic philosophy to falconry. His last work was a treatise on the astrolabe, in which once more he explained "the opinions of the Arabs," this time concerning astronomy. The treatise describes the workings of the astrolabe and its various applications in celestial measurements, using Arabic terms freely and quoting from his other works, particularly his translations of Euclid's *Elements* and the planetary tables of al-Khwarizmi.

Toledo became a center for translation from the Arabic after its recapture from the Moors in 1085 by Alfonso VI, king of Castile and León, the first major triumph of the Reconquista, the Christian reconquest of Al-Andalus. The initiative for the translation movement there seems to have come from Raymond, archbishop of Toledo (1125–51), as evidenced in the dedications of a contemporary Toledan translator, Domenicus Gundissalinus (ca. 1110–ca. 1190).

Gundissalinus, archbishop of Segovia, did several translations and adaptations of Arabic philosophy, including works by al-Kindi, Ibn Rushd, al-Farabi, al-Ghazali, and Ibn Sina, as well as one by the Jewish physician Isaac Israeli. The translations attributed to Gundissalinus were probably done by him in collaboration with others who were fluent in Arabic, though only in one work, the *De Anima* of Ibn Sina, is his name linked with that of a coauthor. There his collaborator was a converted Jew named John ibn David, the Latin Avendehut, who is usually identified with the translator known as John of Seville.

Gundissalinus also wrote five philosophical works on his own, based largely on the books he had translated as well as on Latin sources. He is credited with introducing Arabic-Judaic Neoplatonism to the Latin West and blending it with the Christian Neoplatonism of Saint Augustine and Boethius. His *De Divisione Philosophiae,* which incorporates the systems of both Aristotle and al-Farabi as well as others, is a classification of the sciences transcending the traditional division of studies in the *trivium* (grammar, rhetoric, and logic) and *quadrivium* (arithmetic, geometry, astronomy, and musical theory), and it influenced later schemes of classification.

Plato of Tivoli is known only through his work, at least part of which he wrote in Barcelona between 1132 and 1146. His name appears only as an editor of translations from the Arabic and Hebrew in collaboration with the Jewish mathematician and astronomer Abraham bar Hiyya ha-Nasi, also known as Abraham Judaeus or, in Latin, Savasorda.

Savasorda's most important work is his Hebrew treatise on practical geometry, which he and Plato of Tivoli translated into Latin in 1145 as the *Liber Embadorum.* This was one of the earliest works on Arabic algebra and trigonometry to be published in Latin Europe, and it contains the first solution of the standard quadratic equation to appear in the West. It was also the earliest to deal with Euclid's *On Divisions of Figures,* which has survived not at all in Greek and only partially in Arabic. This work influenced Leonardo Fibonacci, who in his *Practica Geometriae,* written in 1220, devoted an entire section to division of geometrical figures.

Savasorda also collaborated with Plato of Tivoli in translating the *Spherica* by Theodosius of Bithynia, and the two may also have worked together on books by Ptolemy and al-Battani, as well as on Abu Maslama al-Majriti's treatise on the astrolabe. The translations from the Arabic of seven other works are attributed to Plato, with or without Savasorda, five of them astrological, one on divination, and one medical, now lost. One of these works is Ptolemy's great treatise on astrology, the *Tetrabiblos*, which Plato of Tivoli translated into Latin as the *Tetrapartitium*. This was the first Latin translation of Ptolemy, appearing before the *Almagest* and the *Geography*, evidence of the great popularity of astrology in medieval Europe. It has also been suggested that Plato is the author of the Latin translation from the Arabic of Archimedes' *De Mensura Circuli*. Plato's translations were used by both Fibonacci and Albertus Magnus, and printed editions of some of them were published in the late fifteenth and early sixteenth centuries.

Plato's translation of al-Majriti's treatise on the astrolabe is dedicated to "John, son of David." This is probably John of Seville, who in the years 1135–53 translated a score of Arabic works, most of them astrological, but also including an astronomical manual by al-Fargani and a treatise on arithmetic by al-Khwarizmi.

The best known work by John of Seville is his partial translation of the medical section of the pseudo-Aristotelian *Secretum Secretorum* (The Secret of Secrets). This is an apocryphal work that Aristotle is supposed to have composed for Alexander the Great, which a legendary Muslim sage named Ibn Yahya al-Batrik translated from Greek into Chaldean and then into Arabic. A more complete translation was subsequently made by Philip of Tripoli, who in his preface describes how he was in Antioch when he discovered "this pearl of philosophy . . . this book which contains something useful about almost every science."

Translations were also sponsored by Bishop Michael of Tarazona during the years 1119–51, as evidenced by a dedication to him by Hugo Sanctallensis. This appears in Hugo's translation from the Arabic of an abridged version of Ptolemy's *Tetrabiblos*, entitled *Centiloquium*. Hugo's preface says that the *Centiloquium* was commissioned by Michael to serve as a guide to the many astrological works that had been made available to the bishop. Hugo's other translations, all from Arabic sources, are on astrology and various forms of divination, including aeromancy, hydromancy, and pyromancy, prognostication by observing patterns in air, water, and fire, respectively, as well as two short treatises on spatulamancy, foretelling the future by examining the shoulder blades of slaughtered animals.

Gerard of Cremona (1114–1187) was by far the most prolific of all the Latin translators. The few details that are known of Gerard's life come mostly from a short biography and eulogy written by his companions in

Toledo after his death, together with a list of seventy-one works that he had translated. This document was found inserted at the end of Gerard's last translation, which was of Galen's *Tegni* with the commentary of 'Ali ibn Ridwan. It notes that Gerard completed his education in the schools of the Latins before going to Toledo, which he would have reached by 1144 at the latest, when he would have been thirty years old. The vita goes on to say that it was his love of Ptolemy's *Almagest*, which he knew was not available in Latin, that drew Gerard to Toledo, and "there, seeing the abundance of books in Arabic on every subject . . . he learned the Arabic language, in order to be able to translate."

Gerard also lectured on Arabic science, as evidenced by the testimony of the English scholar Daniel of Morley, who had first gone to Paris but had left there in disappointment, traveling to Toledo to hear the "wiser philosophers of the world," as he remarks in his *Philosophia*. Daniel gives a detailed account of meeting "Gerard of Toledo" and listening to his public lectures on Abu Ma'shar's *Great Introduction to the Science of Astrology*. He also listened to lectures by Gallipus Mixtarabe, a Mozarab who collaborated with Gerard on his translation of the *Almagest*, which they seem to have completed in 1175. Otherwise Gerard apparently worked alone, for no collaborators are listed in any of his other translations.

Gerard's translations included Arabic versions of writings by Aristotle, Euclid, Archimedes, Ptolemy, and Galen, as well as works by al-Kindi, al-Khwarizmi, al-Razi, Ibn Sina, Ibn al-Haytham, Thabit ibn Qurra, al-Farghani, al-Farabi, Qusta ibn Luqa, Jabir ibn Hayyan, al-Zarqali, Jabir ibn Aflah, Masha'allah, the Banu Musa, and Abu Ma'shar. The subjects covered in these translations include twenty-four works on medicine; seventeen on geometry, mathematics, optics, weights, and dynamics; fourteen on philosophy and logic; twelve on astronomy and astrology; and seven on alchemy, divination, and geomancy, or predicting the future from geographic features.

Gerard may also have published a number of original works, and several have been tentatively attributed to him, including two glosses on medical texts by Isaac Israeli as well as treatises entitled *Geomantia Astronomica* and *Theorica Planetarium*. However, there is reason to believe that the latter treatise is a work by John of Seville, whose style Gerard adopted in his translations.

More of Arabic science passed to the West through Gerard than from any other source. His translations had a great impact on the development of European science, particularly in medicine, where students in the Latin West took advantage of the more advanced state of medical studies in medieval Islam. His translations in astronomy, physics, and mathematics were also very influential, since they represented a scientific approach to the study of nature rather than the philosophical and theological attitude that had been prevalent in the Latin West.

One of Gerard's contemporaries in Toledo was the Jewish polymath and poet Abraham ibn Ezra (1086–1164), the Latin Avenezra. Ibn Ezra traveled widely, carrying Andalusian Judeo-Muslim culture to Christian Europe, and his visit to London, in 1158–59, helped bring Arabic astrology to England. His Hebrew translation of al-Biruni's *Commentary on the Tables of al-Khwarizmi*, the Arabic original of which is lost, contains interesting information on the influx of Indian ideas into Arabic mathematics and astronomy in the eighth century. Ibn Ezra's own writings include treatises on mathematics, astrology, chronology, and the astrolabe, as well as a work on biblical commentary that was much admired by Spinoza. His astrological works, which number more than fifty, were very popular in medieval Europe and were translated into French, Catalan, and Latin, and later into other languages.

The twelfth-century scholar known as Hermann of Carinthia seems to have learned Arabic in Spain, possibly in Toledo. He is noted for his translation of the Arabic text of Ptolemy's *Planisphaerium*, based on the Arabic text of Abu Maslama al-Majriti. Hermann's translation, which is dated 1143, is the only extant source of the *Planisphaerium*, which treats the problem of mapping circles on the celestial sphere onto a plane, the mathematical basis of the astrolabe. Hermann's other writings include treatises on the astrolabe and astrology, a commentary on Euclid and other mathematical works, and a translation of the astronomical tables of al-Khwarizmi.

Hermann did several of his translations in collaboration with the English scholar Robert of Chester, a younger contemporary of Adelard's; they worked together at several places in southern France and Spain, including Toledo. Robert's solo translations include al-Khwarizmi's *Algebra* (dated Segovia, 1145); a treatise on the astrolabe (London, 1147); a set of astronomical tables for the longitude of London (1149–50), based upon the tables of al-Zarqali and al-Battani; and a revision, also for the meridian of London, of Adelard's version of the tables of al-Khwarizmi.

Robert also translated *De Compositione Alchemie* by Romanus Morienus, one of the earliest works on alchemy rendered into Latin, dated 1144. This is an apocryphal work supposedly written by Morienus, a Christian hermit in Jerusalem, to whom the "secrets of all divinity" had been revealed by a mystic named Adfar of Alexandria, who had found and mastered the astrological works of Hermes Trismegistus, the legendary sage.

One of the extant manuscripts of Robert's revisions of al-Khwarizmi's work contains astronomical tables for the longitude of Hereford in England, dated 1178, which have been attributed to the twelfth-century English astronomer Roger of Hereford. Roger wrote several works on astronomy and astrology in the decade 1170–80. One of these, *Liber de Divisione Astronomiae*, begins with the phrase "In the name of God the pious and merciful," the traditional opening of an Islamic treatise, suggesting that this is a translation from the Arabic, though the author is unknown.

Alfred of Sareshel, another twelfth-century English scholar, dedicated one of his translations to Roger of Hereford. Alfred did translations of several Aristotelian works from Arabic, together with commentaries, and he also translated the sections on geology and alchemy from Ibn Rushd's *Kitab al-Shifa*, which he entitled *De Mineralibus*. Alfred seems to have learned Arabic in Spain, where he probably did his translation of Ibn Rushd, and he also appears to have used Greek sources, particularly in his works on Aristotle, whose natural philosophy and metaphysics he introduced to England.

The most important interface between the Greek, Latin, and Arabic cultures in the twelfth century was the Norman realm in southern Italy and Sicily, the "Kingdom of the Two Sicilies." The Normans had driven the Byzantines from their last footholds in southern Italy and then subdued the Saracens in Sicily. When Count Roger I conquered Palermo in 1091, it had been under Muslim domination for nearly two centuries. He reduced the Muslims to the status of serfs except in Palermo, his capital, where he employed the most talented of them as civil servants, so that Greek, Latin, and Arabic were spoken in the Norman court and used in royal charters and registers. Under his son Roger II (r. 1130–54), Palermo became a center of culture for both Christians and Muslims, surpassed only by Cordoba and Toledo. Beginning under Roger II, and continuing with his successors, the Sicilian court sponsored numerous translations from both Greek and Arabic into Latin.

Roger II was particularly interested in geography, but he was dissatisfied with the existing Greek and Arabic geographical works. Thus in 1138 he wrote to al-Idrisi (1100–1166), the distinguished Muslim geographer and cartographer, who was then living in Cueta, and invited him to visit Palermo, saying, "If you live among the Muslims, their kings will contrive to kill you, but if you stay with me you will be safe." Al-Idrisi accepted the offer and lived in Palermo until Roger's death in 1154, after which he returned to Cueta and passed his remaining days there.

Roger commissioned al-Idrisi to create a large circular relief map of the world in silver, the data for which came from Greek and Arabic sources, principally Ptolemy's *Geography*, as well as from travelers and the king's envoys. The silver map has long since vanished, but its features were probably reproduced in the sectional maps in al-Idrisi's Arabic geographical compendium, *Kitab Nuzhat al-Mushtaq fi Ikhtiraq al-Afar*, which has survived. The compendium deals with both physical and descriptive geography, with information on political, economic, and social conditions in the lands around the Mediterranean and in the Middle East, and is thus a veritable encyclopedia of the medieval world. Al-Idrisi's work was a popular textbook in Europe for several centuries, and a number of abridgements were done, the first in Rome in 1592. A Latin translation was published in Paris in 1619, and a two-volume French translation was done in 1830–40, entitled *Géographie d'Edrisi*.

Frederick II of Hohenstauffen (r. 1212–50), the Holy Roman emperor and king of the Two Sicilies, was a grandson of the emperor Frederick I Barbarossa and the Norman king Roger II. Known in his time as *stupor mundi*, "the wonder of the world," he had been raised from age seven to twelve in Palermo, where he grew up speaking Arabic and Sicilian as well as learning Latin and Greek. When he became emperor in 1211, at the age of fourteen, he turned away from his northern dominions to his Kingdom of the Two Sicilies, where, like his Norman predecessors, who were known as "baptized sultans," he indulged himself in his harem in the style of an Oriental potentate.

Frederick was deeply interested in science and mathematics, and he invited a number of scholars to his brilliant court, most notably John of Palermo, Master Theodorus, and Michael Scot, calling them his "philosophers." He subsidized their scientific writings and translations, which included Aristotle's works on physics and logic, some of which he presented in 1232 to the professors at Bologna University. The letter Frederick sent with the gift told of how he had loved learning since his youth, and of how he still took time from affairs of state to read in his library, where numerous manuscripts of all kinds "classified in order, enrich our cupboards."

Frederick's scholarship is evident in his famous book on falconry, *De Arte Venandi cum Avibus* (The Art of Hunting with Birds). This is a scientific work on ornithology as well as a detailed and beautifully illustrated manual of falconry as an art rather than a sport. Frederick acknowledged his debt to Aristotle's *Zoology*, which had been translated by Michael Scot in the thirteenth century. But he was critical of some aspects of the work, as he writes in the preface to his manual: "We have followed Aristotle when it was opportune, but in many cases, especially in that which regards the nature of some birds, he appears to have departed from the truth. That is why we have not always followed the prince of philosophers, because rarely, or never, had he the experience of falconing which we have loved and practiced always."

One of those with whom Frederick corresponded was the renowned mathematician Leonardo Fibonacci (ca. 1170–after 1240), who had been presented to him when he held court at Pisa in about 1225. Leonardo had at that time just completed his treatise on squared numbers, the *Liber Quadratorum*, which he dedicated to Frederick, noting, "I have heard from the Podesta of Pisa that it pleases you from time to time to hear subtle reasoning in Geometry and Arithmetic."

Leonardo, who was born in Pisa, writes about his life in the preface to his most famous work, the book on calculations entitled *Liber Abbaci*. His father, a secretary of the Republic of Pisa, was in around 1192 appointed director of the Pisan trading colony in the Algerian city of Bugia (now Bejaïa). Leonardo was brought to Bugia by his father to be trained in the art of calculating, which he learned to do "with the new Indian numerals," the so-called Hindu-Arabic numbers, which he would introduce to Europe in his *Liber Abbaci*. His father also sent him on business trips to Provence, Sicily, Egypt, Syria, and Constantinople, where he met with Latin, Greek, and Arabic mathematicians. In around 1200 he returned to Pisa, where he spent the rest of his days writing the mathematical treatises that made him the greatest mathematician of the Middle Ages.

The five works of Leonardo's that have survived are the *Liber Abbaci*, first published in 1202 and revised in 1228; the *Practica Geometriae* (1220–21), on applied geometry; a treatise entitled *Flos* (1225), sent to Frederick II in response to mathematical questions that had been put to Leonardo by John of Palermo at the time of the emperor's visit to Pisa; an undated letter to Master Theodorus, one of the "philosophers" in the court of Frederick II; and the *Liber Quadratorum* (1225). This last work contains the famous "rabbit problem": "How many pairs of rabbits will be produced in a year, beginning with a single pair, if in every month each pair produces a new pair which become productive from the second month on?" The solution to this problem gave rise to the so-called Fibonacci numbers, a progression in which each number is the sum of

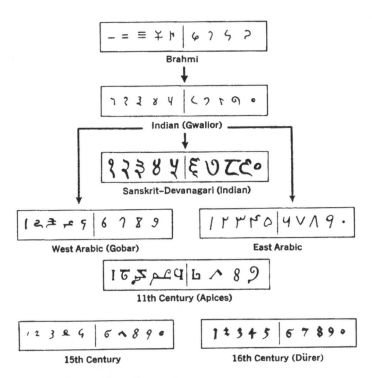

Brahmi

Indian (Gwalior)

Sanskrit–Devanagari (Indian)

West Arabic (Gobar)

East Arabic

11th Century (Apices)

15th Century

16th Century (Dürer)

Development of the Hindu-Arabic numbers.

the two that precede it (e.g., 1, 1, 2, 3, 5, 8, 13, 21 . . .), a mathematical wonder that continues to fascinate mathematicians. Leonardo's sources, where they can be traced, include Greek, Roman, Indian, and Arabic works; by synthesizing them and adding to them with his own creative genius, he stimulated the beginning of the new European mathematics.

Leonardo dedicated his *Flos* to John of Palermo, whom he also mentions in the introduction to the *Liber Quadratorum*. John's only known work is a Latin translation of an Arabic treatise on the hyperbola, which may be derived from a work by Ibn al-Haytham on the same subject.

Master Theodorus, who is usually referred to as "the Philosopher," was born in Antioch. He served Frederick as secretary, ambassador, astrologer, and translator, from both Greek and Arabic into Latin, and he was also the emperor's chief confectioner. One of his works is a translation of an Arabic work on falconry. He served the emperor until the time of his death, in around 1250, when Frederick regranted the estate that "the late Theodore our philosopher held so long as he lived."

Theodorus had probably succeeded Michael Scot as court astrologer. Michael was born in the last years of the twelfth century, probably in Scotland, though he might possibly be Irish. Nothing is known of his university studies, but his references to Paris indicate that he may have studied and lectured there as well as in Bologna, where he did some medical research in 1220 or 1221. He may have learned Arabic and some Hebrew in Toledo, where in about 1217 he translated al-Bitruji's *On the Sphere*, with the help of Abuteus Levita, a Jew who later converted to Christianity. By 1220 he had completed what became the standard Latin version of Aristotle's *On Animals,* from a ninth-century Arabic version by al-Bitriq, as well as the *De Caelo* and the *De Anima* with Ibn Rushd's commentaries. He had become a priest by 1224, when Pope Honorius II appointed him as archbishop of Cashel, in Ireland, and obtained benefices for him in England. He declined the appointment as archbishop, saying that he did not speak Irish, and was then given further benefices in England and Scotland by the archbishop of Canterbury.

When Leonardo Fibonacci completed his revised version of *Liber Abbaci* in 1228 he sent it to Michael, who by that time seems to have entered the service of Frederick II as court astrologer. Michael wrote for the emperor a Latin summary of Ibn Rushd's *De Animalibus* as well as a voluminous treatise known in English as *Introduction to Astrology*. The latter work covers every aspect of astrology and divination, including necromancy, or conjuring up the spirits of the dead to reveal the future or influence the course of coming events, as well as nigromancy, or black magic, dealing with dark things performed by night rather than by day.

Michael condemned necromancy and black magic, but he delighted in telling stories of nigromancers and magicians. He says that the best nigromancer in France was a clerk named Gilbertus, presumably Gerbert d'Aurillac, whom he says conjured up demons who explained to him the use of the astrolabe and the principles of astronomy, after which he reformed and became bishop of Ravenna and then pope.

Frederick addressed a long series of extraordinary questions to Michael, who inserted the questionnaire as an addendum to a work entitled *Libers Particularis*. Frederick's interest in necromancy is indicated in one of the questions he asked Michael: "And how is it that the soul of a living man which has passed away to another life than ours cannot be induced to return by first love or even by hate, just as it had been nothing, nor does it seem to care at all for what it has left behind whether it be saved or lost." Michael boasted that he could answer all of the questions asked by the emperor, including his query as to "whether one soul in the next world knows another and whether one can return to this life to speak and show one's self; and how many are the pains of hell."

All of this led to Michael's posthumous fame as a magician, clouding his reputation as a scientist and translator, which is in any event controversial. Roger Bacon referred to Michael as "a notable inquirer into matter, motion, and the course of the constellations," but at the same time he listed him among those translators who "understood neither sciences nor languages, not even Latin," and said that his translations were for the most part done by a Jew named Andrew. Bacon credits Michael with having introduced the natural philosophy of Aristotle to the Latin West, though Michael actually transmitted only *De Animalibus*.

Along with Gerbert d'Aurillac, Michael was said to have sold his soul to the devil in exchange for his knowledge of the black arts and the magic of science. Dante writes of him in Canto XX of the *Inferno*, where he is pointed out in the fourth ditch of the eighth circle of Hell, among the other diviners: "That other, round the loins / So slender of his shape, was Michael Scot / Practised in every slight of magic wile."

Michael's reputation as a magician of the black arts endured until modern times, particularly in Scotland, where his skill in the black arts was celebrated in a doggerel rhyme:

> *A wizard, of such dreaded fame,*
> *That when, in Salamanca's cave,*
> *Him listed his magic wand to wave,*
> *The bells would ring in Notre Dame!*

PARIS AND OXFORD I:
REINTERPRETING ARISTOTLE

By the beginning of the thirteenth century European science was on the rise, stimulated by the enormous influx of Greco-Arabic works translated into Latin and used at the new universities that sprang up all over Europe, supplanting the cathedral schools of the earlier medieval era.

The earliest of these institutions of higher learning was the university of Bologna, founded in 1088, followed in turn by those of Paris (ca. 1150), Oxford (1167), Salerno (1173, a refounding of the medical school), Palenzia (ca. 1178), Reggio (1188), Vicenza (1204), Cambridge (1209), Salamanca (1218), and Padua (1222), to name only the first ten, with another ten founded in the remaining years of the thirteenth century. Twenty-five more were founded in the fourteenth century, and another thirty-five in the fifteenth, so that by 1500 there were eighty universities in Europe, evidence of the tremendous intellectual revival that had taken place in the West, beginning with the initial acquisition of Greco-Arabic learning in the twelfth century.

Bologna became the archetype for later universities in southern Europe, Paris and Oxford for those in the northern part of the continent. Bologna was renowned for the study of law and medicine, Paris for logic and theology, and Oxford for philosophy and natural science. Training in medicine was based primarily on the teachings of Hippocrates and Galen, while studies in logic, philosophy, and science were based on the works of Aristotle and commentaries upon them, at first translated from Arabic and then later from Greek.

Although Aristotle's works formed the basis for most nonmedical studies at the new universities, some of his ideas in natural philosophy, particularly as interpreted in commentaries by Averroës (Ibn Rushd), were strongly opposed by Catholic theologians. One point of objection to Aristotle was his notion that the universe was eternal, which denied the act of God's creation; another was the determinism of his doctrine of cause and effect, which left no room for divine intervention or other miracles. Still another objection was that Aristotle's natural philosophy was pantheistic, identifying God with nature, which derived from the Neoplatonic interpretation of Aristotelianism by Avicenna (Ibn Sina).

This led to a decree, issued by a council of bishops in Paris in 1210, forbidding the teaching of Aristotle's natural philosophy in the university's faculty of arts. The ban was renewed in 1231 by Pope Gregory IX, who issued a bull declaring that Aristotle's works on natural philosophy were not to be read at the University of Paris "until they shall have been examined and purged from all heresy." The ban seems to have remained in effect for less than a quarter century, for a list of texts used at the University of Paris in 1255 includes all of Aristotle's available works.

The controversy was renewed in 1270 when the Bishop of Paris, Etienne Tempier, condemned thirteen propositions derived from the philosophy of Aristotle or from Aristotelian commentaries by Averroës. This gave rise to the doctrine of "double-truth," in which an idea might be declared true if demonstrated by reason in physics and metaphysics, while a contradictory concept could be independently considered true in theology and the realm of faith. Pope John XXI, after seeking the advice of theologians, issued a bull in 1277 in which he condemned 219 propositions, including the original thirteen listed by Tempier, threatening excommunication of anyone who held even a single erroneous doctrine. That same year a similar condemnation was issued by Tempier as well as by the archbishop of Canterbury, Robert Kilwardby, whose edict was renewed in 1284 by his successor, John Pecham. A number of the propositions were declared to be erroneous because their determinism placed limits on the power of God.

Meanwhile European scholars were absorbing the Greco-Arabic learning that they had acquired and used to develop a new philosophy of nature, which although primarily based upon Aristotelianism differed from some of Aristotle's doctrines right from the beginning. The leading figure in the rise of the new European philosophy of nature was Robert Grosseteste (ca. 1168–1253). Born of humble parentage in Suffolk, England, he was educated at the cathedral school at Lincoln and then at the University of Oxford. He taught at Oxford and went on to take a master's degree in theology, probably at the University of Paris. He was then appointed chancellor of the University of Oxford, where he probably also lectured on theology, while beginning his own study of Greek. When the first Franciscan monks came to Oxford in 1224 Grosseteste was appointed as their reader. He finally left the university in 1235 when he was appointed bishop of Lincoln, his jurisdiction including Oxford and its schools.

Grosseteste's works are divided into two periods, the first when he was chancellor of Oxford and the second when he was bishop of Lincoln. His writings in the first period include his commentaries on Aristotle and the Bible and most of his independent treatises, while those in the second period are principally his translations from the Greek: Aristotle's *Nichomachean Ethics* and *On the Heavens*, the latter along with his version of the commentary by Simplicius, as well as several theological works.

Grosseteste's commentaries on Aristotle's *Posterior Analytics* and *Physics* were among the first and most influential interpretations of those works. These two commentaries also presented his theory of science, which he put into practice in his own writings, including six works on astronomy and one on calendar reform, as well as treatises entitled *The Generation of the Stars, Sound, The Impressions of the Elements, Comets, The Heat of the Sun, Color, The Rainbow,* and *The Tides,* in which he attributed tidal action to the moon.

Grosseteste was the first medieval scholar to deal with the methodology of science, which for him involved two distinct steps. The first of these was a combination of deduction and induction, which he termed "composition" and "resolution," a method for arriving at definitions. As Grosseseste put it: "This method involves two procedures, one being by composition and the other by resolution. Aristotle teaches first the method of arriving at the definition by composition, because this method is like a progression from the more universal and simple to the more composite. The method of resolution is the opposite of that."

The second step was what Grosseteste called verification and falsification, a process necessary to distinguish the true cause from other possible causes. He based his use of verification and falsification on two assumptions about the nature of physical reality. The first of these was the principle of the uniformity of nature, in support of which he quoted Aristotle's statement that "the same cause, provided that it remains in the same condition, cannot produce anything but the same effect." The second was the principle of economy, which holds that the best explanation is the simplest, that is, the one with the fewest assumptions, other circumstances being equal. Here again he quoted Aristotle, who said that power from natural agents proceeds in a straight line "because nature operates in the shortest way possible." Beginning with these assumptions, Grosseteste's method was to distinguish between possible causes "by experience and reason," rejecting theories that contradicted either factual evidence or an established theory verified by experience.

Grosseteste believed that it was impossible to understand the physical world without mathematics. He noted that "the science that is concerned with the study of radiant lines and figures [i.e., optics] falls under geometry . . . ; the science of constructing machines, as under architecture and other mechanical arts, falls under the science of the figures of bodies; the science of harmonies falls under arithmetic; and the science which sailors use to direct the course of ships by the appearance of the stars is subordinate to geometry."

The use of mathematics made it essential to perform measurements that resulted in a number, though in doing so there was an inescapable inaccuracy, which made all human measurements conventional, as opposed to the certitude of geometry. But although geometry, for example, could give the "reason for the fact," in the sense of describing a phenomenon in optics such as reflection of light, it could not provide the physical causes involved. Thus a complete explanation of optical phenomena requires not only geometry, but a knowledge of the physical nature of light that causes it to move as it does in being reflected by a mirror, in which the angle of incidence equals the angle of reflection.

Grosseteste believed that the study of optics was the key to an understanding of the physical world, and this gave rise to his Neoplatonic "Metaphysics of Light." He believed that light is the fundamental corporeal substance of material things and produces their spatial dimensions, as well as being the first principle of motion and efficient causation. According to his optical theory, light travels in a straight line through the propagation of a series of waves or pulses, and because of its rectilinear motion it can be described geometrically. This was similar to the acoustical theory he presented in his commentary on Aristotle's *Posterior Analytics*, where he writes that "when the sounding body is struck and vibrating, a similar vibration and similar motion must take place in the surrounding contiguous air, and this generation progresses in every direction in straight lines."

Grosseteste thought that the same theory, which he called the "multiplication of species," could be used to explain the propagation of any disturbance, be it light, sound, heat, mechanical action, or even astrological influence. Thus the study of light was of crucial importance for an understanding of nature. He also believed that light, by which he meant not only visible radiation but the divine emanation as well, was the means by which God created the universe, and that through it soul and body interacted in man.

The study of optics was divided by Grosseteste into three parts: phenomena involving vision, mirrors (catoptrics, or reflection), and lenses (dioptrics, or refraction). He discussed the third part more fully than the other two, noting that it had been "untouched and unknown among us until the present time," and suggested applications of refraction that in the seventeenth century would be realized through the invention of the telescope and the microscope. "This part of optics," he wrote, "when well understood, shows us how we may make things a very long distance off appear as if placed very close . . . and how we may make small things placed at a distance appear any size we want, so that it may be possible for us to read the smallest letters at incredible distances, or to count sand, or grains, or seeds, or any sort of minute objects."

Grosseteste developed a theory of refraction in an attempt to explain the focusing of light by a "burning-glass," or spherical lens. An experiment would have shown him that his law of refraction was incorrect, but apparently he never put his law to the test, although it was one of the basic tenets of his scientific method that if a theory was contradicted by observation it must be abandoned.

Grosseteste's application of his scientific method is evident in his treatise *The Rainbow*, in which he broke with Aristotelian theory by holding that the phenomenon was due to refracted rather than reflected light. Although his theory was incorrect, he posed the problem in such a way that investigations by those who followed after him approached closer to the true solution through criticizing his efforts. His work on the rainbow inspired some verses written about 1270 by the French poet Jean de Meun in his continuation of Guillaume de Lorris's *Romance of the Rose.* These are in chapter 83, where "Nature explains the influence of the heavens," telling of how the clouds, "to give solace to the earth"

> *Are wont to bear, ready at hand, a bow*
> *Or two or even three if they prefer,*
> *The which celestial arcs are rainbows called,*
> *Regarding which nobody can explain,*
> *Unless he teaches optics in some school,*
> *How they are varicolored by the sun,*
> *How many and what sorts of hues they show,*
> *Wherefore so many and such different kinds,*
> *Or why they are displayed in such a form.*

There is little in Grosseteste's writing to indicate that he was a Christian bishop, but in his treatise *On the Fixity of Motion and Time* he differed from the Aristotelian doctrine that says the universe is eternal, for that contradicted his belief in God's creation. His Christian beliefs are also evident in another treatise, *On the Order of the Emanation of Things Caused from God,* in which he said that he wished men would cease questioning the biblical account of the Creation.

Grosseteste also wrote a number of treatises on astronomy. The most important of these was *De Sphaera,* in which he discussed elements of both Aristotelian and Ptolemaic theoretical astronomy. He also wrote of Aristotelian and Ptolemaic astronomy in his treatise on calendar reform, *Compotus correctorius,* where he used Ptolemy's system of eccentrics and epicycles to compute the paths of the planets, though he noted, "These modes of celestial motion are possible, according to Aristotle, only in the imagination, and are impossible in nature, because according to him all nine spheres are concentric." Grosseteste also wrote of astrological influences in his treatise *On Prognostication,* but he later condemned astrology, calling it a fraud and a delusion of Satan's.

Grosseteste's *De Sphaera* was written at about the same time as a treatise of the same name by his contemporary John of Holywood, better known by his Latin name, Johannes de Sacrobosco. Little is known of his life other than the facts that he became a monk at the Augustine monastery of Holywood, and that after studying at Oxford he was admitted in 1221 to the University of Paris, where he was elected professor of mathematics.

Sacrobosco's principal extant works are three elementary textbooks on mathematics and astronomy: *De Sphaera, De Computo Eccliastico,* and *De Algorismo,* all of which are frequently bound together in the same manuscript. Sacrobosco's fame is principally due to his *De Sphaera,* an astronomy text based on Ptolemy and his Arabic commentators, most notably al-Farghani. The text was first used at the University of Paris and then at all schools throughout Europe, and it continued in use until the late seventeenth century. His *De Computo Eccliastico* points out the errors in the Julian calendar, proposing a solution very similar to the reform adopted by Pope Gregory XIII three and a half centuries later. His *De Algorismo,* which taught the techniques of calculating with positive integers, was the most widely used manual of arithmetic in the medieval era, continuing in use until the sixteenth century.

Grosseteste's efforts in framing a new philosophy of nature were continued by Albertus Magnus (ca. 1200–1280). Born to a family of the military nobility in Bavaria, he studied liberal arts at the University of Padua, where he was recruited into the Dominican order by its master general, Jordanus of Saxony. Albertus then studied theology and taught in Germany before enrolling in the University of Paris circa 1241, where he lectured on theology for seven years before he was sent to open a school in Cologne. His students included Thomas Aquinas, who came from Italy to study with him, either in Paris or Cologne. Albertus was appointed provincial of the German Dominicans in 1253, and in 1260 he became bishop of Regensburg, a post that he resigned two years later, after which he spent the rest of his life preaching and teaching.

Albertus played a crucial role in rediscovering Aristotle and making his philosophy of nature acceptable to the Christian West. The main problem involved in the Christian acceptance of Aristotle was the conflict between faith and reason, particularly in the Averroist interpretation of Aristotelian thinking, with its determinism and its view of the eternity of the cosmos. Albertus sought to resolve this conflict by regarding Aristotle as a guide to reason rather than an absolute authority, and declaring that where his ideas conflicted with either revealed religion or observation, he must be wrong. Albertus held that natural philosophy and theology often spoke of the same thing in different ways, and so he assigned to each of them its own realm and methodology, assured that there could be no contradiction between reason and revelation.

Albertus undertook the task of interpreting Aristotle at the request of his Dominican brethren, who wished to understand the Aristotelian worldview; he explains this in the prologue to his commentary on Aristotle's *Physics*, where he says that his purpose is "to make all parts of philosophy intelligent to the Latins."

The most original contributions made by Albertus were in botany and the life sciences, where his work was distinguished by his acute observations and skill in classification. His attitude toward scientific method is evident in his commentary on the pseudo-Aristotelian *De Plantis*, the principal source of botanical knowledge until the sixteenth century, where in discussing the native plants known to him he writes, "In this sixth book we will satisfy the curiosity of the students rather than philosophy. . . . Syllogisms cannot be made about particular natures, of which experience (*experimentum*) alone gives certainty."

Although Albertus was very modern in his scientific thinking, he was still medieval in his views on such matters as magic, divination, and astrology. He writes in his *Summa Theologica* of his belief that magic is due to demons: "For the saints expressly say so, and it is the common opinion of all persons, and it is taught in that part of necromancy which deals with images and rings and mirrors of Venus and seals of demons." Albertus writes of astrology in almost all of his scientific treatises, describing the effects produced by such celestial phenomena as conjunctions of the planets, to which he attributes "great accidents and great prodigies and a general change of the state of the elements and of the world."

Ulrich Engelbert of Strasburg, a pupil of Albertus's, describes him as "a man in every science so divine that he may well be called the wonder and miracle of our time." Thomas Aquinas writes of him with equal admiration, saying, "What wonder that a man of such whole-hearted devotion and piety should show superhuman attainments in science."

Albertus was canonized by Pope Pius XI on 16 December 1931, and ten years later Pope Pius XII declared him the patron saint of all those who cultivate the natural sciences.

Thomas Aquinas (ca. 1225–1274) was born near Monte Cassino in southern Italy, where his father served the emperor Frederick II in his war against the papacy. He began his education at the Benedictine abbey of Monte Cassino, after which he went to the newly founded University of Naples, where he was introduced to the works of Aristotle. After joining the Dominicans, he was sent for further studies to Cologne and then Paris, where his teachers included Albertus Magnus.

Aquinas spent two periods as professor at the University of Paris, 1256–59 and 1269–72, and in the interim he was associated in turn with the papal courts of Alexander IV, Urban IV, and Clement IV. After his second professorship at Paris he returned to Naples to start a Dominican school, which he directed until a few months before his death in 1274. He was canonized by Pope John XXII on 18 July 1323, and subsequently he was presented by the Roman Catholic Church as the most representative teacher of its doctrines. His writings are still taught at Catholic universities.

Aquinas, like Albertus Magnus, tried to resolve the conflict between theology and natural science and show that there could be no real contradiction between revelation and reason. Arguing against those who said that natural philosophy was contrary to the Christian faith, he writes in his treatise *Faith, Reason and Theology* that "even though the natural light of the human mind is inadequate to make known what is revealed by faith, nevertheless what is divinely taught to us by faith cannot be contrary to what we are endowed with by nature. One or the other would have to be false, and since we have both of them from God, he would be the cause of our error, which is impossible."

This involved him in disputes at the University of Paris in 1268–70 over the Aristotelian philosophy that he and Albertus had introduced. The condemnation of Averroist doctrines by the bishop of Paris in 1270 may have been directed against some of the teachings of Aquinas, one being that the creation of the world cannot be demonstrated by reason alone. This and other interpretations by Aquinas were his solution to the problem of adapting Aristotelianism to Christian theology, creating the philosophical system that came to be called Thomism.

The lengths to which Aquinas went can be seen in his attempt to fit the biblical account of the Ascension into the Aristotelian cosmos. According to Ephesians 4:10, Christ "ascended up far beyond all heavens, that he might fill all things," which presented problems for Aquinas as he tried to square this with Aristotle's philosophy and his model of the homocentric crystalline spheres.

Meanwhile, in the thirteenth century, translations were still being made from Arabic into Latin. Some of these were done under the patronage of King Alfonso X (1221–84) of Castile and León, known in Spanish as el Sabio, or the Wise. Alfonso's active interest in science led him to sponsor translations of Arabic works in astronomy and astrology, including a new edition of the *Toledan Tables* of the eleventh-century Cordoban astronomer al-Zarqali. This edition, known as the *Alfonsine Tables*, included some new observations but retained the Ptolemaic system of eccentrics and epicycles.

Grosseteste's most renowned disciple was Roger Bacon (ca. 1219–92), who acquired his interest in natural philosophy and mathematics while studying at Oxford. He received an M.A. either at Oxford or Paris, in around 1240, after which he lectured at the University of Paris on various works of Aristotle's. He returned to Oxford circa 1247, when he met Grosseteste and became a member of his circle.

Bacon became a Franciscan monk in about 1257, and soon afterward he experienced difficulties, probably because of a decree restricting the publication of works outside the order without prior approval. In any event, Pope Clement IV issued a papal mandate on 22 June 1266 asking Bacon for a copy of his philosophical writings. The mandate ordered Bacon not only to send his book but to state "what remedies you think should be applied in these matters which you recently intimated were of such great importance," and "to do this without delay as secretly as you can."

Bacon eventually replied with three works—*Opus Maius, Opus Minus,* and *Opus Tertium*—along with a letter proposing a reform of learning in the Catholic Church. He maintained that there were two types of experience, one obtained through mystical inspiration and the other through the senses, assisted by instruments and quantified in mathematics. The program of study he recommended included languages, mathematics, optics, experimental science, and alchemy, followed by metaphysics and moral philosophy, which, under the guidance of theology, would lead to an understanding of nature and through that to knowledge of the Creator.

Within the next few years Bacon wrote three more works, the *Communia Naturalium, Communia Mathematica,* and *Compendium Studii Philosophie,* the last of which castigated the Franciscan and Dominican orders for their educational practices. Sometime between 1277 and 1279 he was condemned and imprisoned in Paris by the Franciscans, possibly because of their censure of heretical Averroist ideas. Nothing further is known of his life until 1292, when he wrote his last work, the *Compendium Studii Theologii.*

Bacon appropriated much of Grosseteste's concept of the "Metaphysics of Light" with its "multiplication of species," as well as his mentor's emphasis on mathematics, particularly geometry. In his *Opus Maius* Bacon states that "in the things of the world, as regards their efficient and generating causes, nothing can be known without the power of geometry"; he also says, "Every multiplication is either according to lines, or angles or figures." His ideas on optics also repeat those of Grosseteste. But he does go beyond Grosseteste in his commentary on Alhazen (Ibn al-Haytham), particularly his theory of the eye as a spherical lens, basing his own anatomical descriptions on those of Hunayn ibn Ishaq and Avicenna (Ibn Sina).

Bacon clearly states his scientific method in Part VI of the *Opus Maius,* "De Scientia Experimentali," which also derives from Grosseteste. There he writes of the "three great prerogatives" of experimental science, the first being "that it investigates by experiment the noble conclusions of all the sciences." The second prerogative, according to Bacon, is that experiment adds new knowledge to existing sciences, and the third is that it creates entirely new areas of science. He also emphasizes the vital importance of mathematics in science, writing that "no science can be known without mathematics."

Bacon used his scientific method to study the rainbow, where he improved on Grosseteste's theory in his understanding that the phenomenon was due to the action of individual raindrops, though he erred in rejecting refraction as part of the process.

Other works by Bacon include the *Epistola de Secretis Operibus Artis et Naturae et de Nullitate Magiae,* which describes wonderful machines such as self-powered ships, automobiles, airplanes, and submarines.

> Machines for navigation can be made without rowers so that the largest ships on rivers or seas will be moved by a single man in charge with greater velocity than if they were full of men. Also cars can be made so that without animals they will move with unbelievable rapidity.... Also flying machines can be constructed so that a man sits in the midst of the machine revolving some engine by which artificial wings are made to flap like a flying bird.... Also a machine can easily be made for walking in the sea and rivers, even to the bottom without danger.

On another topic, Bacon writes that "it has been proved by certain experiments" that life can be greatly extended by "secret experiences." One of his recommendations for achieving an exceptionally long life involves eating the specially prepared flesh of flying dragons, which he says also "inspires the intellect," or so he was told "without deceit or doubt from men of proved trustworthiness."

Writings such as this gave Bacon the posthumous reputation of being a magician and diviner who had learned his black arts from Satan. Early in the seventeenth century a book was published in London entitled *The famous historie of Fryer Bacon, containing the wonderful things that he did in his life, also the manner of his death, with the lives and deaths of the two conjurers, Bungey and Vandermast.* The book purports to tell the story of Bacon's life and magical exploits, including his creation of a brazen head that could talk and foretell the future and protect England from her enemies. After fashioning the "talking head" Bacon and Bungey waited for it to speak, but nothing happened for three weeks. Then "after some noyse the head spake these two words. 'TIME IS'; and again after an interval, 'TIME WAS'; and again, 'TIME IS PAST,' and therewith fell downe, and presently followed a terrible noyse, with strange flashes of fire."

11

PARIS AND OXFORD II: THE EMERGENCE OF EUROPEAN SCIENCE

The reinterpretation of Aristotle in the thirteenth century led to the emergence of European science, particularly through the researches of Robert Grosseteste and his followers in Paris and Oxford.

One of the pioneers of this new European science was Jordanus Nemorarius (fl. ca. 1220), a contemporary of Grosseteste's. Virtually nothing can be said about the life of Jordanus, who is known only through the inclusion of his works in the *Bibliomania*, a catalog of the library of Richard de Fournival done sometime between 1246 and 1260, in which twelve treatises are ascribed to him.

Jordanus made his greatest contribution in the medieval "science of weights" (*scientia de ponderibus*), now known as "statics," the study of forces in equilibrium. One of the concepts he introduced was that of "positional gravity" (*gravitas secundum situm*), which he expressed in the statement that "weight is heavier positionally, when, at a given position, its path of descent is less oblique." An example would be a block on an inclined plane, whose apparent weight, the force with which it presses against the surface, is greater if the angle of inclination is less. This is equivalent to resolving the weight into two components, one perpendicular to the plane, which is the apparent weight or "positional gravity," and the other parallel to the surface.

The concept of positional gravity was used by Jordanus in his study of the most basic problem in statics, that of the beam balance, where two weights are suspended on either side of a fulcrum. According to Jordanus, "if the arms of the balance are unequal, then if equal weights are suspended from their extremities, the balance will be depressed on the side of the longer arm." His proof of this statement again uses the concept of positional gravity, which in this case is equal to the weight of the object times its lever arm, the perpendicular distance from the fulcrum to the line of action of the weight. This is now known as the "moment of a force," or "torque," a measure of its effectiveness in rotating the balance, equilibrium resulting if the two torques are equal and opposite.

Jordanus then proceeded to "prove" the law of the lever, which is that two objects will balance each other if their weights are inversely proportional to their lever arms. Here he used the concept of "work," the product of the weight of an object times the distance through which it is lifted or otherwise moved, the first clear definition of this fundamental concept in physics. He also introduced the concept of "virtual velocity," that is, one that is vanishingly small, since a real movement cannot take place in a system under equilibrium. He used this concept in examining two objects balanced on a lever, where in a virtual displacement the positive work done in lifting one weight is equal to the negative work in lowering the other, leading to the conclusion that the system is in equilibrium. His proof involves what is known as the "axiom of Jordanus," which says that the motive power that can lift a given weight a certain height can lift a weight k times heavier to $\frac{1}{k}$ times that height, where k is any number.

The same concepts were used by Jordanus in studying the equilibrium of two different connected weights on inclined planes of different inclinations, which he treated as a generalized case of the law of the lever. The proof refers to a triangle ABC which has BC as its base and a right angle at A, where a pulley connects two weights, with $w(1)$ on side AB and $w(2)$ on side AC. He showed that the two weights will be in equilibrium if their positional gravities are equal—that is, if the components of each weight down its plane are equal to that of the other in the opposite direction. This can be reduced to the equation $w(1)/w(2) = AB/AC$, which is equivalent to the law of the lever.

Jordanus also made contributions in mathematics, where he shows no trace of Islamic influence, following rather in the Greco-Roman tradition of Nichomachus and Boethius. He was the first to use the letters of the alphabet in arithmetical problems for greater generality, and he presented algebraic problems leading to linear and quadratic equations. He also worked in geometry and, following Archimedes, solved problems involving the determination of the center of gravity of triangles and other plane figures, as well as doing pioneering research in stereographic projection.

The scholar known as Gerard of Brussels, who appears to have been associated with Jordanus, may have been the first European to deal with kinematics, a purely mathematical description of motion. His treatise on kinematics, *De Motu*, written sometime between 1187 and 1260, was strongly influenced by Euclid and Archimedes.

During the second quarter of the fourteenth century a group of scholars at Merton College, Oxford, developed the conceptual framework and technical vocabulary of the new science of motion. These were Thomas Bradwardine, William Heytesbury, John of Dumbleton, and Richard Swineshead, who continued the Oxford tradition in science initiated by Robert Grosseteste.

Thomas Bradwardine (ca. 1290–1349) received his bachelor's, master's, and doctoral degrees at Oxford in the years 1321–48, and from 1323 until 1335 he was a fellow of Merton College. In 1339 he was chaplain and perhaps confessor to King Edward III, and he accompanied the king to France in the campaign of 1346. He was elected archbishop of Canterbury on 4 June 1349, but he died of the plague on 26 August of that same year.

Bradwardine's principal work is the *Tractatus Proportionum*, completed in 1328. The problem that Bradwardine tried to solve in this work was to find a suitable mathematical function for the Aristotelian law of motion, which states that the velocity (v) of an object is proportional to the power (p) of the mover divided by the resistance (r) of the medium. Bradwardine focused on the change in velocity rather than the velocity itself and tried to show how this was related to power and resistance. Stated mathematically, he looked for a functional relation between the dependent variable v and the two independent variables p and r; that is, given values for p and r, to find the corresponding value of v. After trying and rejecting a number of equations, he finally settled on a law of motion that, in modern terminology, states that the velocity is proportional to the logarithm of p/r. Bradwardine never tested his law of motion, which would have shown him that it was not correct. Nevertheless, his formulation of the problem in terms of a mathematical functional relationship was an important step forward in the science of dynamics, one that was followed by his successors at both Oxford and Paris. Their researches established the foundations of the late medieval tradition of the *calculatores*, those who studied the quantitative variation of motion, power, and qualities in space and time.

William Heytesbury's name, variously spelled, appears in the records of Merton College for 1330 and 1338–39, and he may be the William Heighterbury or Hetisbury who was chancellor of the university in 1371. His most influential work is his *Regulae Solvendi Sophismata*, dating from 1335.

Heytesbury, in his *Regulae*, defined uniform acceleration as motion in which the velocity is changing at a constant rate, either increasing or decreasing. For such motion he defined acceleration as the change in velocity in a given time, which would be negative in the case of deceleration. He also introduced the notion of instantaneous velocity—that is, the speed at a particular moment—defining it as the distance traveled by a body in a given time if it continued to move with the speed that it had at that moment. He showed that, for uniformly accelerated motion, the average velocity during a time interval is equal to the instantaneous velocity at the midpoint of that interval. This was known as the mean speed rule of Merton College; it was adopted by Heytesbury's successors at both Oxford and Paris.

John of Dumbleton (fl. 1331–49) is mentioned as a fellow of Merton College in the years 1338–48. His best-known writing is the *Summa Logicae et Philosophiae Naturalis*, a vast work that critically discusses most of the topics in physics and philosophy of his time. Here he covered rates of change, including motion, change of quality, and growth, in relation to a fixed scale such as distance or time. A change was said to be uniform when there were equal variations in equal intervals of time, and "difform" when the variations increased or decreased in time. Thus in uniform motion equal distances are covered in equal intervals of time, while in difform motion, the intervals traversed in successive time intervals increase or decrease.

Richard Swineshead, who was known as the "Calculator," is best remembered as the author of the *Liber Calculationum* (ca. 1340–50), a work that became famous for its extensive use of mathematics in physics. The *Liber Calculationum* concentrates on calculating the values of physical variables and solving problems about their changes. The work is divided into sixteen treatises, of which the last three are devoted to "Local Motion"; there he elaborates at exhausting length on Bradwardine's law of dynamics in every conceivable type of motion, many of which have no known parallel in nature. The *Liber Calculationum* was disseminated widely in Europe, and it was printed at Padua (ca. 1477), Pavia (1498), and Venice (1520). The Venetian edition was later transcribed for the great German mathematician and philosopher Leibniz (1646–1716), who praised Swineshead for having introduced mathematics into scholastic philosophy.

Advances in the theory of motion were also being made in Paris, beginning with the work of Jean Buridan (ca. 1295–ca. 1358). Little is known of Buridan's origins other than the fact that he was born in the diocese of Arras. Soon after 1320 he obtained his master's degree at the University of Paris, where he was twice elected rector, first in 1328 and again in 1340.

Buridan's extant writings consist of the lectures he gave at the University of Paris, where the curriculum was based largely on the study of Aristotle, along with textbooks on logic, grammar, mathematics, and astronomy. Buridan wrote his own textbook on logic as well as two advanced treatises on the subject. All of his other writings are commentaries and books about Aristotle's principal works.

Buridan's philosophy of science is enunciated in his *Questions* on Aristotle's *Physics* and *Metaphysics*. There he makes the distinction between premises whose necessity is determined through logic and those based on empirical evidence, whose necessity is conditional "on the supposition of the common cause of nature." He held that the principles of natural science are of the second type, noting that they "are not immediately evident . . . but they are accepted because they have been observed to be true in many instances and to be false in none."

Buridan's most important contribution to science is his so-called impetus theory, the revival of a concept first proposed in the sixth century by John Philoponus. He explains the continued motion of a projectile as being due to the impetus it received from the force of projection, and says it "would endure forever if it were not diminished and corrupted by an opposing resistance or something tending to an opposed motion." Buridan defines impetus as a function of the body's "quantity of matter" and its velocity, which is equivalent to the modern concept of momentum, or mass times velocity, where mass is the inertial property of matter, its resistance to a change in its state of motion. As applied to the case of free fall, Buridan explains that gravity not only is the primary cause of the motion but also imparts additional increments of impetus to the body as it falls, thus accelerating it—that is, increasing its velocity.

Buridan extended his concept of impetus to explain the motion of the celestial spheres, which in Aristotelian cosmology rotated with constant velocity. He argued that there was no need to have immaterial "intelligences" as the unmoved movers of the celestial spheres, as supposed by Aristotle, because their motion was inertial after the initial impetus they received from the Creator. "For it could be said that God, in creating the world, set each celestial orb in motion ... and, in setting them in motion, he gave them an impetus capable of keeping them in motion without there being any need of his moving them any more." He added that this was why God could rest on the seventh day after the Creation, for the inertia of the celestial spheres would keep them in motion without the need for any additional divine effort.

In one of the questions in *De Caelo et Mundo*, Buridan asks if a proof can be given for Aristotle's geocentric model, in which the earth is at rest at the center of the cosmos with the stars and other celestial bodies rotating around it. He notes that many in his time believed the contrary, that the earth is rotating on its axis and that the stellar sphere is at rest, adding that it is "indisputably true that if the facts were as this theory supposes, everything in the heavens would appear to us just as it now appears." In support of the earth's rotation, he says that it is better to account for appearances by the simplest theory, and it is more reasonable to think that the vastly greater stellar sphere is at rest and the earth is moving, rather than the other way around. But, after refuting the usual arguments against the earth's rotation, Buridan says that he himself believes the contrary, using the argument that a projectile fired directly upward will fall back to its starting point, which is true, at least approximately, whether or not the earth is rotating.

The impetus theory was adopted by Buridan's students and became known throughout Europe, though in a corrupted form that restored some Aristotelian notions. Aside from that, Buridan is credited with eliminating explanations involving Aristotelian final causes from physics. His books were required reading at universities until the seventeenth century and would have been read by both Copernicus and Galileo. Copernicus used some of Buridan's arguments in discussing the earth's motions, and Galileo revived the theory of impetus in formulating his own laws of kinematics and dynamics.

Like other famous medieval scholars, Buridan was the subject of apocryphal stories. One of these tales, perpetuated by the poet François Villon, tells of how Buridan had an affair with the wife of Charles V of France, who had him tied up in a sack and thrown into the Seine.

The most distinguished of his students was Nicole Oresme (ca. 1320–1382), who studied under Buridan at the University of Paris in the 1340s. He was chosen grand master of the university's College of Navarre in 1356 and three years later became secretary of the dauphin of France, the future King Charles V. From about 1369 he was employed by Charles to translate some of Aristotle's Latin texts into French, for which he was rewarded in 1377 when, at the behest of the king, he was made bishop of Lisieux, a post he held until his death in 1382.

Oresme gave a graphical demonstration of the Merton mean speed rule in his *Tractus de Configurationibus Qualitatum et Motuum*, written in the 1350s while he was at the College of Navarre in Paris. The graph plots the velocity (v) on the vertical axis as a function of the time (t) on the horizontal axis, as, for example, in the case of a body starting from rest and accelerating so that its velocity increases by 2 feet per second every second. Graphing the motion for 4 seconds, we see that the velocity increases each second, from 0 to 2 then 4 then 6 then 8 feet per second at the end of 4 seconds. This graphs in the form of a straight line rising from 0 to 8 feet per second over a time interval of 4 seconds, forming a right triangle with a height of 8 and a base of 4. The acceleration (a) is equal to the slope of the straight line, which is $\frac{8}{4}$, or 2, in units of feet per second per second. The average velocity is half of the final velocity, which is $\frac{8}{2}$, or 4 feet per second. The mean speed rule then gives the distance s traveled in 4 seconds as 4 times 4, or 16 feet. The rule can be applied over each one-second interval, so that the average velocity for each second increases from 1 to 3 to 5 to 7 feet per second. The distance in feet traveled in the first second is then 1, in the second 3, in the third 5, and in the fourth 7. These results can be generalized by the equations $v = a \times t$, and $s = a/2 \times t^2$. These are the kinematic equations formulated by Galileo in his *Dialogue Concerning the Two New Sciences* (1638), where he used Oresme's demonstration of the Merton mean speed rule in his proof.

Oresme also had original ideas in astronomy, which he presented in his *Livre du Ciel et du Monde d'Aristote*, written in 1377 for Charles V. One of these was his comparison of the eternal motion of the celestial spheres to a perpetual mechanical clock, set in motion by God at the moment of their creation. He writes that "it is not impossible that the heavens are moved by a power or corporeal quality in it, without violence and without work, because the resistance in the heavens does not incline them to any other movement nor to rest but only that they are not moved more quickly."

Orseme objected to the Aristotelian notion that the earth was the stationary center of the finite cosmos and the reference point for all motion and gravitation. He argued that motion, gravity, and the directions in space must be regarded as relative, saying that God, through his omnipotence, could create an infinite space and as many universes as he chose. Oresme was thus able to reject the idea that the earth was the fixed center of the cosmos to which all gravitational motions were directed. Instead he proposed the idea that gravity was simply the tendency of bodies to move toward the center of spherical mass distributions. Gravitational motion was relative only to a particular universe; there was no absolute direction of gravity applying to all of space.

Oresme proposed, "subject to correction, that the earth is moving with daily motion and the heavens not. And first I will declare that it is impossible to show the contrary by any observation; secondly from reason; and thirdly I will give reasons in favor of the opinion."

Oresme's arguments in favor of the earth's motion would later be used by both Copernicus and Galileo. Despite all of these arguments Orseme, who at the time had just been appointed bishop of Lisieux, in the end rejected the idea of the diurnal rotation of the earth as being contrary to Christian doctrine, saying, "For God fixed the earth, so that it does not move, notwithstanding the reasons to the contrary."

Orseme's attitude was not uncommon among his clerical contemporaries, for in his position as bishop he was sworn to uphold the doctrines of the Catholic Church, even when they conflicted with his own philosophical ideas.

Meanwhile, advances were being made in other areas of science, namely, natural philosophy, cosmology, magnetism, astronomy, and optics, as well as in mathematics and its application to astronomy and other fields of science.

Giles of Rome (ca. 1247–1316) was a student of Thomas Aquinas's in Paris. When Etienne Tempier delivered his second condemnation of Aristotelianism and Averroism in 1277, Giles's writings were censured and he was forced to leave Paris. He returned to Paris in 1285 at the request of Pope Honorius IV, after having retracted several of his theses. He was appointed archbishop of Bourges in 1295 by Pope Boniface VIII, and died in 1316 during a stay at the papal court in Avignon.

One of the original ideas proposed by Giles was that there are natural minima below which physical substances cannot exist, thus implying an atomic theory of matter. He investigated the nature of the vacuum through experiments with a clepsydra, a cupping glass, and a siphon, showing that the void exerted a force of suction. He disagreed with Aristotle and his own contemporaries in holding that celestial matter is identical to that of the terrestrial world. He rejected Aristotle's model of homocentric spheres in favor of Ptolemy's theory of eccentrics and epicycles, saying that observational evidence must settle the controversy between the Aristotelian "physicists" and the Ptolemaic "mathematicians." He also admitted the possibility of a plurality of worlds.

The earliest extant treatise on magnetism is by Peter Peregrinus, of whom virtually nothing is known except what appears in his work and in possible references to him by Roger Bacon. Peter's treatise—actually, a letter—is the *Epistola Petri Peregrini de Maricourt ad Sygerum de Foucaucourt, Miltem, De Magnete* (Letter on the Magnet of Peter Peregrinus of Maricourt to Sygerus of Foucaucourt, Soldier). Peter concluded the letter with the note that it had been "Completed in camp, at the siege of Lucera, in the year of our Lord 1269, eighth day of August." This would indicate that Peter was at the time in the army of Charles of Anjou, king of Sicily, who was then besieging the city of Lucera in southern Italy.

The *Epistola* is in two parts, of which the first, in ten chapters, describes the properties of the lodestone, or magnetic rock; the second is devoted to the construction of three instruments using magnets. Peter's observations led him to make the distinction between the north and south magnetic poles; to establish the rules for the attraction and repulsion of magnetic poles; to show the magnetization of iron by bringing it in contact with a magnet; and to demonstrate that a magnetic needle when broken in half forms two separate magnets. He showed that a magnetic needle oriented itself in the north-south direction, thereby inventing the compass, which he said could be used to map the meridians of the earth's magnetic field. He believed, mistakenly, that the poles of a magnetic needle point to the poles of the celestial sphere, the points about which the stars appear to be rotating, which are actually projections of the earth's axis of rotation. He attempted to construct a perpetual motion machine using magnets, and he blamed his failure on his lack of skill rather than the impossibility of creating an eternal source of energy. The *Epistola* was very popular in the late medieval era, as evidenced by the fact that there are at least thirty-one extant manuscript copies. It had a great impact on William Gilbert, who in his famous *De Magnete* (1600), paid tribute to Peter Peregrinus and acknowledged his debt to his predecessor.

Campanus of Novara, who flourished in the second half of the thirteenth century, is best known for his translation of Euclid's *Elements*, but he was also a distinguished astronomer. Little is known of his life other than that he was chaplain to Popes Urban IV, Nicholas IV, and Boniface VIII, and that he spent his last years at the Augustine monastery at Viterbo, in Italy, where he died in 1296.

Campanus's principal astronomical work is his *Theorica Planetarium*, a description of the structure and dimensions of the universe according to Ptolemaic theory, together with instructions for the construction of an instrument, later called an equatorium, designed to find the position of any of the celestial bodies at a given time. Campanus based his calculations on the work of the ninth-century Arabic astronomer Alfraganus (al-Farghani), who had in turn derived them from the *Planetary Hypotheses* of Ptolemy. Campanus probably also learned about the equatorium from an Arabic source, for descriptions of equatoria were written nearly two centuries earlier in Al-Andalus by Ibn al-Samh and Ibn al-Zarqali.

One of the most notable of the early European astronomers was William of St. Cloud, who flourished in France during the late thirteenth century. The earliest date of his activity is 28 December 1285, when he observed a conjunction of Saturn and Jupiter, to which he refers in his *Almanach*, completed in 1292. His other major work is his *Calendrier de la Reine*, also completed in 1292, dedicated to Queen Marie of Brabant, widow of Philip III, "the Bold," and which he translated into French at the request of Jeanne of Navarre, wife of Philip IV, "the Fair."

Queen Marie's *Calendrier* represents William's effort to put the calendar on a purely astronomical basis. This led him to contradict the computations in the ecclesiastical calendar, which he found full of errors, indicating the need for calendar reform. The purpose of his *Almanach* was to provide listings in which the positions of the celestial bodies were given directly, as contrasted to earlier tables, which only gave the elements by which those positions could be calculated. He points out the errors in the earlier planetary tables and shows how he corrected them. These tables were those of Toledo, used in the Muslim calendar, and of Toulouse, the adaptation of the *Toledan Tables* to the Christian calendar. William's *Almanach* makes no mention of the *Alphonsine Tables*, which were not used in Paris before 1320.

William's observations were remarkably precise, evidence of the high level that had been achieved in European astronomy by his time. By comparing his astronomical observations with ancient Greek values he was able to measure the change in the spring equinox, which he interpreted as a steady precession rather than the trepidation theory introduced by Theon of Alexandria.

A new set of astronomical tables was completed in 1327 by John of Ligneres and his students, whose work was heavily dependent on the *Toledan Tables*. This work, known as the *Large Tables*, included a catalog of the positions of the forty-seven brightest stars and was easier to use than any of the earlier tables. Thus it became very popular, though it was eventually supplanted in Paris by the *Alphonsine Tables*.

Levi ben Gerson (1288–1344) was a polymath who wrote books on astronomy, physics, mathematics, and philosophy, as well as commentaries on the Bible and the Talmud. He lived in Orange and Avignon, which were not affected by the expulsion of the Jews from France in 1306 by King Philip the Fair. He was also on good terms with the papal court in Avignon, as evidenced by his dedication of one of his works to Pope Clement IV in 1342.

Levi's greatest work is his *Milhamot Adonai* (The Wars of the Lord), a philosophical treatise in six books, the fifth of which is devoted to astronomy. Here Levi presents his model of the universe, based on several Arabic sources, principally al-Battani, Jabir ibn Aflah, and Ibn Rushd. His model differed in important respects from that of Ptolemy, whose theories did not always agree with observations made by Levi. This was particularly so in the case of Mars, where Ptolemy's theory had the apparent size of the planet varying by a factor of six, while Levi's observation found that it only doubled. The instruments used by Levi included one of his own invention, the "Jacob's staff," a device to measure angles in astronomical observations. He also employed the camera obscura—invented by Alhazen (Ibn al-Haytham)—for observing eclipses and determining the eccentricity of the sun's orbit. Levi's astronomical work was influential in Europe for five centuries, and his Jacob's staff was also used for maritime navigation until the mideighteenth century.

During the first half of the fourteenth century an important school of astronomy was active at Oxford. The most notable of the Oxford astronomers was Richard of Wallingford (ca. 1292–1336). Richard, the son of a blacksmith, studied at Oxford from about 1308 until 1315, when he joined the Benedictine order at St. Albans abbey. Two years later he was sent back to Oxford for further studies, remaining there until 1327, when he was appointed abbot of St. Albans. After his appointment he visited Avignon for the papal confirmation, and when he returned to St. Albans he found that he had contracted leprosy; he died from the disease in 1336.

Richard's works were all written during his years at Oxford. One of these was his *Quadripartitium*, the first comprehensive treatise on spherical trigonometry written in Latin Europe. Richard's most important work was his *Tractus Albionis*, which dealt with the theory, construction, and use of an instrument that he had invented called the Albion, a form of equatorium used to perform all sorts of astronomical measurements and calculations.

Richard also built an enormous mechanical clock, ten feet in diameter, which he installed on the wall of the south transept in the abbey church. Besides the time of day, it also showed the motion of the celestial bodies, the phases of the moon, and the tides. The clock was destroyed in the sixteenth century, but several drafts of Richard's design have survived, the oldest extant plans of any mechanical clock, the most sophisticated chronometer of the medieval era. (The earliest mechanical clocks in Europe appear to date from the end of the thirteenth century.) The mechanical clock led to the notion of time as a physical quantity that could be expressed numerically in terms of units on a scale and used in scientific theories. Since Richard's clock was also a planetarium, it lent credence to the notion that the universe was a divinely designed clockwork.

The earliest appearance of a mechanical clock in literature seems to be in a passage of Dante's *Paradiso,* in the last lines of Canto X, written between 1316 and 1321, a decade or so before Richard built his clock.

> *Forthwith*
> *As clock, that calleth up the spouse of God*
> *To win her Bridegroom's love at matin's hour.*
> *Each part of other fitly drawn and urged,*
> *Sends out a tinkling sound, of note so sweet,*
> *Affection springs in well-disposed breast;*
> *Thus saw I move the glorious wheel; thus heard*
> *Voice answering voice, so musical and soft,*
> *It can be known but where day endless shines.*

Another area in which the new European science developed was optics, the study of light, which had begun at Oxford with the work of Robert Grosseteste and his disciple Roger Bacon. The first significant advance beyond what they had done was by the Polish scholar Witelo (b. ca. 1230–35; d. after ca. 1275).

Witelo's best-known work is the *Perspectiva,* which is based on the works of Robert Grosseteste and Roger Bacon as well as those of Alhazen, Ptolemy, and Hero of Alexandria. It would seem that the *Perspectiva* was not written before 1270, since it makes use of Hero's *Catoptrica,* the translation of which was completed by William of Moerbeke (b. ca. 1220–35; d. before 1286) on 31 December 1269.

Witelo adopted the "metaphysics of light" directly from Grosseteste and Bacon, and in the preface to the *Perspectiva* he says that visible light is simply an example of the propagation of the power that is the basis of all natural causes. But he disagrees with Grosseteste and Bacon where they say that light rays travel from the observer's eye to the visible object, and instead follows Alhazen in holding that the rays emanate from the object to interact with the eye.

The *Perspectiva* describes experiments performed by Witelo in his study of refraction. Here his method is similar to that of Ptolemy; he measures the angle of refraction for light in passing from air into glass and also into water, for angles of incidence ranging from 10 to 80 degrees. He tried to explain the results by a number of mathematical generalizations, attempting to relate the differences in refraction to the difference in the densities of the two media. In addition, he produced the colors of the spectrum by passing light through a hexagonal crystal, observing that the blue rays were refracted more than the red.

Witelo also studied refraction in lenses, making use of the concept later known as the principle of minimum path. He justified this by the metaphysical notion of economy, saying, "It would be futile for anything to take place by longer lines, when it could better and more certainly take place by shorter lines."

Witelo followed Grosseteste in holding that the "multiplication of species" could be used to explain the propagation of any effect, including the divine emanation and astrological influences. In the preface to the *Perspectiva*, which he addresses to William of Moerbeke, he writes "of corporeal influences sensible light is the medium." He also writes that "there is something wonderful in the way in which the influence of divine power flows in to things of the lower world passing through the powers of the higher world."

The next advances in optics were made by Dietrich of Freiburg (ca. 1250–ca. 1311), who is sometimes called Theodoric. Dietrich, who is thought to be from Freiburg in Saxony, entered the Dominican order and probably taught in Germany before studying at the University of Paris, circa 1275–77.

Dietrich's principal work is his treatise *On the Rainbow and Radiant Impressions*, the latter term meaning phenomena produced in the upper atmosphere by radiation from the sun or any other celestial body. He was the first to realize that the rainbow is due to the individual drops of rain rather than the cloud as a whole. This led him to make observations with a glass bowl filled with water, which he used as a model raindrop, for, he writes, "a globe of water can be thought of, not as a diminutive

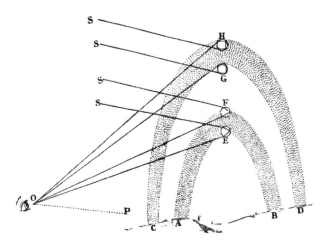

The formation of primary and secondary rainbows, from Newton's *Opticks*.

spherical cloud, but as a magnified raindrop." His observations and geometrical analysis led him to conclude that light is refracted when it enters and leaves each raindrop, and that it is internally reflected once in creating the primary bow and twice for the secondary arc, the second reflection reversing the order of the colors in the spectrum. Although he made a number of errors in his analysis, his theory was far superior to those of any of his predecessors, and it paved the way for researches by his successors.

Dietrich's theory of the rainbow is very similar to that of his Persian contemporary Kamal al-Din al-Farisi. Dietrich does not cite the work of al-Farisi, but since it was never translated from Arabic into Latin he was probably not aware of it. In any event, it seems that the emerging European science had by the beginning of the fourteenth century reached a level comparable to that of Arabic scientific research, at least in optics. But whereas the work of al-Farisi was the last great achievement of Arabic optics, Dietrich's researches would be an important stage in the further development of European studies in the science of light, culminating in the first correct theories of the rainbow and other optical phenomena in the seventeenth century.

12

FROM BYZANTIUM TO ITALY: GREEK INTO LATIN

T he only medieval state that retained an unbroken cultural link with the classical Greek world was the Byzantine Empire, with its capital at Constantinople, ancient Byzantium. The link was very tenuous at times, as when the empire was almost overrun by invaders, losing much of its territory in both Europe and Asia. Constantinople was captured and sacked in 1204 by the army of the Fourth Crusade and the Venetian fleet, after which the Byzantine Empire was reduced to several small states, two of them in Asia Minor. One of these, ruled by the Lascarid dynasty, had its capital at Nicaea, while the Comneni dynasty reigned from its capital at Trebizond. In 1261 the Greeks of Nicaea recaptured Constantinople, which once again became capital of the Byzantine Empire, a state that was much reduced in size and power compared to what it had been in the reign of Justinian.

During the century prior to the Fourth Crusade several Italian city-states had obtained commercial concessions in the Byzantine Empire, allowing them to set up trading establishments in Constantinople and other cities of the empire. Genoa, Venice, Pisa, and Amalfi had concessions in Constantinople, where they built docks and warehouses along the Golden Horn as well as residences and churches for merchants and their families. These Latin concessions continued in existence until the end of the empire, serving as centers not only for trade but also for cultural interaction between the Greek East and the Latin West.

One important instance of such a cultural interchange occurred in 1136, during the reign of the Byzantine emperor John II Comnenus, when the Holy Roman Emperor Lothair sent a mission to Constantinople to discuss theological differences between the Roman Catholic and Greek Orthodox churches. The mission was headed by Anselm, bishop of Havelberg and later archbishop of Ravenna. After Anselm arrived in Constantinople he and his entourage went to the Pisan quarter on the Golden Horn to discus theology with a group of Greek clerics headed by Nicetas, archbishop of Nicomedia. Anselm writes,

> There were present not a few Latins, among them three wise men skilled in the two languages [Latin and Greek] and most learned in letters, mostly James a Venetian, Burgundio a Pisan, and the third, most famous among Greeks and Latins above all others for his knowledge of both literatures, Moses by name, an Italian from the city of Bergamo, and he was chosen by all to be an interpreter for both sides.

The first of these scholars, James of Venice, is known in Latin as Iacobus Veneticus Grecus, which could mean that he was a member of the Greek community of Venice. In any event, he was fluent in both Greek and Latin, as indicated by an entry for the year 1128 in the chronicle of Robert of Torigni, abbot at Mont-Saint-Michel, who writes that "James, a clerk of Venice, translated from Greek into Latin certain books of Aristotle and commented upon them, namely the *Topics,* the *Prior* and *Posterior Analytics,* and the [*Sophistici*] *Elenchi,* although there was an older version of these books."

James was the first European scholar in the twelfth century to introduce the works of Aristotle to the Latin West. Besides the works mentioned by Robert of Torigni, James was the first to translate from the Greek Aristotle's *Physica, De Anima, Metaphysica,* and *Parva Naturalia.* His commentary on the *Sophistici Elenchi* shows that he was aware of Byzantine scholarship on this subject in Constantinople, which was an unrivaled source for the works of Aristotle and other Greek writers. James's translations, together with their revisions, formed the basis for much of Aristotelian studies in Europe until the sixteenth century.

Burgundio the Pisan traveled frequently from Pisa to Constantinople and to Sicily, another rich source of Greek manuscripts. His translations from the Greek also included the *Aphorisms* of Hippocrates and ten works of Galen's, as well as Aristotle's *Meteorology*.

Moses of Bergamo, whom Anselm mentions as the interpreter in the theological disputation of 1136, lived at that time in the Venetian quarter of Constantinople. He writes in one of his letters that he learned Greek so that he could translate previously unknown manuscripts into Latin. He spent years collecting Greek manuscripts, for which he paid a total of three pounds of gold, he says, but they were all destroyed in a fire in 1130. The only translations of his that have survived are works of Greek grammarians and early Byzantine theologians.

Translations from Greek to Latin were also done in Sicily during the reign of William I (r. 1154–66), the son and successor of Roger II; he continued his father's patronage of learning. The two principal translators during his reign were Henricus Aristippus and Eugene the Emir, both of them members of the royal administration who left eulogies of William, commemorating him as a philosopher-king who opened his court to the world's leading scholars. Aristippus became archbishop of Catania in 1156 and four years later was placed in charge of the entire administration of the Sicilian kingdom. He was the first to translate from the Greek two of Plato's dialogues, *Meno* and *Phaedo*, as well as the fourth book of Aristotle's *Meteorology*, works that remained in use until the early Renaissance. Aristippus also served as envoy to the court of Manuel II Comnenus in Constantinople, where the emperor presented him with a beautiful codex of Ptolemy's *Almagest* as a present to King William. The first Latin translation of this manuscript from Greek to Latin was made in Palermo by an anonymous visiting scholar circa 1160. Other works translated from Greek to Latin at the Sicilian court by this scholar include Euclid's *Optica* and *Catoptrica*, Proclus's *De Motu*, and Hero of Alexandria's *Pneumatica*.

The unknown scholar who translated these works notes that in doing so he received considerable assistance from Eugene the Emir, "a man most learned in Greek and Arabic and not ignorant of Latin." Eugene, who held the Arabic title of emir in the royal administration, was probably a Greek, as evidenced by his surviving poetry.

The Dominican monk William of Moerbeke, in Belgium, was the most prolific of all medieval scholars who translated from Greek into

Latin. Moerbeke is known to have visited Nicaea in the spring of 1260, when the Byzantines still had their capital there, not recapturing Constantinople until the following year, and he may very well have acquired Greek manuscripts at that time. He took part in the Second Council of Lyons (May–June 1274), whose goal was to bring about a reunion between the Greek and Latin churches, and at a pontifical mass he sang the *Credo* in Greek together with Byzantine clerics.

Thomas Aquinas is said to have suggested to Moerbeke that he complete the translation of Aristotle's works directly from the Greek. Moerbeke says that he took on this task "in spite of the hard work and tediousness which it involves, in order to provide Latin scholars with new material for study."

Moerbeke's Greek translations included the writings of Aristotle, commentaries on Aristotle, and works by Archimedes, Proclus, Hero of Alexandria, Ptolemy, and Galen. The popularity of Moerbeke's work is evidenced by the number of extant copies of his translations, including manuscripts from the thirteenth to fifteenth centuries, printed editions from the fifteenth century onward, versions in English, French, Spanish, and even modern Greek done from the fourteenth century through the twentieth. His translations led to a better knowledge of the actual Greek texts of several works, and in a few cases they are the only evidence of lost Greek texts, such as that of Hero's *Catoptrica*.

Moerbeke's only original work is a treatise on divination entitled *Geomantia*, which was evidently quite popular, as demonstrated by the several extant Latin manuscripts and a French translation done in 1347. Witelo, in the dedication to Moerbeke in his *Perspectiva*, praises him for his "occult" inquiry into the influence of divine power on humans.

Another Western scholar who visited Constantinople in search of ancient Greek manuscripts was Peter of Abano (1250–ca. 1313), who while there found works by Aristotle, Dioscorides, and Galen, among others. His translations from the Greek include a volume of Aristotle's *Problems*, the first Latin translation of this work; Aristotle's *Rhetoric*; Dioscorides's *De Materia Medica*; and six treatises of Galen's.

The most famous of Peter's original works is his *Conciliator Differentiarum Philosophorum et Praecipue Medicorum*, which he completed in 1303, while he was teaching at the University of Paris. In this enormous tome Peter tries to reconcile the conflicting views of the medical writers and philosophers who had preceded him. The *Conciliator* comprises more than two hundred questions, or "differences," which Peter says he and his colleagues had been debating for the past decade. The first and eighteenth questions, for example, concern the differences of opinion about whether the heart is the center of the human nervous system, as Aristotle holds, or whether it is the brain, as Peter says. His conclusion concerning the first question is that "the regulative power of the body resides in the brain"; regarding the eighteenth, he says that "the brain is the seat of sensation and emotion." Question 67 asks, "Is life possible below the equator?" It seems that this question occurred to Peter after he met Marco Polo, who had returned to Venice in 1295 after his celebrated journey to the Far East.

Another well-known work by Peter is his *Lucidator Dubitabilium Astronomiae*, in which he discusses disputed doctrines in astronomy and astrology. Here he suggests that the stars are not fixed in the outermost celestial sphere, as Aristotle has it in his model of the cosmos, but move freely in space, an entirely new idea that would become part of modern cosmology. A number of passages in this work indicate that Peter associated spirits and intelligences with the celestial bodies, one of which he describes as "perpetual and incorruptible, leading through all eternity a life most sufficient unto itself, nor ever growing old."

Peter's writings on astrology and other occult sciences gave him the reputation of being a magician. Gabriel Naude, in 1625, called Peter "a man who appeared as a prodigy and miracle in his age . . . he was the greatest magician of his age and learned the seven liberal arts from seven familiar spirits whom he held captive in a crystal." But Naude went on to say that Peter had in later life abandoned "the idle curiosity of his youth to devote himself wholly to philosophy, medicine and astrology."

Meanwhile Byzantium was undergoing a cultural revival under the Palaeologues, the dynasty that ruled the Byzantine Empire after its recapture of Constantinople from the Latins in 1261 until its fall to the Turks in 1453. Michael VIII Palaeologus (r. 1259–82), the founder of the dynasty, reestablished the University of Constantinople, which had been closed during the Latin occupation.

The first head of the reopened university was George Acropolites, who lectured on the mathematics of Euclid and Nicomachus as well as the philosophy of Aristotle. Acropolites had been a student of Nicephorus Blemmydes (ca. 1197–1272), who wrote a handbook on physics, astronomy, and logic as well as a geographical synopsis and several theological commentaries, all of which were used at the University of Constantinople. Another student of Blemmydes' who taught at the university was George Pachymeres (1242–1310), who was primarily a historian but who also wrote on mathematics and the theory of music.

A number of scholars from Byzantium went west on diplomatic missions or to teach at European universities. One of the first of the diplomat scholars was the monk Maximos Planudes (1260–1310), who was sent by Andronicus II (r. 1282–1328) on an embassy to Venice in 1296. After his return to Constantinople Planudes translated from Latin to Greek a number of books that he had acquired in Italy, including works by Cicero, Ovid, Saint Augustine, and Boethius. The vast number of extant manuscripts of his translations show that they were often used as texts for teaching Greek at universities in Italy, which began with the emergence of Italian humanism and its interest in reviving classical letters. He also wrote two works on mathematics, one of them a commentary on Diophantus's *Arithmetica* and the other a treatise on the "Hindu" numbers being used by Arabic mathematics, the first mention of this system of numeration by a Byzantine scholar.

Italian humanism had focused on Latin learning until the time of Petrarch (1304–74) and his pupil Boccaccio (1313–75), who set out to learn Greek so that they could read Homer and Plato and other Greek writers in the original. At Avignon Petrarch began to study Greek with Barlaam of Calabria (d. 1348), who had been sent to the papal court on a diplomatic mission by the emperor Andronicus III (r. 1328–41). But Barlaam died soon afterward and Petrarch never learned proper Greek.

The renowned Byzantine humanist Manuel Chrysoloras (ca. 1350–1417) was sent on a diplomatic mission from Constantinople to Venice in 1390. During his stay in Venice, which ended the following year, Chrysoloras was offered a contract to give Greek lessons in Florence, where he taught from 1396 until 1400. The leading Florentine statesmen and humanists who flocked to his lectures included the chancellor Leonardo Bruni, who is quoted as saying that Chrysoloras inspired him to learn a language that no Italian had understood for seven hundred years.

The textbook that Chrysoloras wrote for his lessons, the *Erotemata* (Questions), was translated into Latin by Guarino of Verona (1374–1460), who studied Greek with him in Constantinople from 1403 to 1408. Guarino went on to teach Greek in Venice, Florence, and Ferrara, where he remained for the last thirty years of his life, attracting students from as far away as England. He was also noted for his translations, which included works by Homer, Herodotus, Plutarch, and Strabo, whose *Geography* he translated in his latter years.

Chrysoloras was contemporary with the Florentine humanist Poggio Bracciolini, who in 1417 discovered the full text of Lucretius's *De Rerum Natura*, with its philosophy of nature based on the atomic theory of Democritus. This was printed later in the fifteenth century and reprinted many times thereafter, reviving interest in the atomic theory, which had been in obscurity since antiquity.

By the beginning of the fourteenth century Byzantium was flowering in a last renaissance, even as the empire was being engulfed by the rising tide of the Ottoman Turks. The Ottomans established their first capital in 1326 at Bursa, in northwestern Asia Minor, and crossed over into Europe twenty years later, when they captured the city of Callipolis (Gallipoli), on the Dardanelles. The Turkish capture of Callipolis was recorded by the Byzantine scholar Demetrius Cydones, renowned for his translations from Latin into Greek, most notably the *Summa Theologica* of Thomas Aquinas. According to Cydones, the fall of Callipolis began an exodus of Greeks from Byzantium to Italy, Spain, and even farther "towards the sea beyond the Pillars [the Strait of Gibraltar]."

The extent of the Palaeologan cultural revival is revealed by the beautiful mosaics and frescoes that have survived in the church of St. Savior in Chora, now known as Kariye Camii, in Istanbul, built in the years 1315–21 by Theodore Metochites (ca. 1250–1332), prime minister under Andronicus II.

Metochites was one of the most renowned figures of his time in Byzantium, the head of the government as well as a distinguished theologian, philosopher, astronomer, mathematician, physicist, poet, and patron of the arts, a leader in the artistic and intellectual renaissance in the early Palaeologan era. His student and protégé Nicephorus Gregoras wrote of him in admiration, "From morning to evening he was most wholly and eagerly devoted to public affairs as if scholarship was absolutely indifferent to him; but later in the evening, having left the palace, he became absorbed in science to such a degree as if he were a scholar with absolutely no connection with any other affairs."

Metochites' most important scientific work is his *Introduction to Astronomy*, which is based on Ptolemy's *Almagest*. He made no astronomical observations, merely recalculating the old parameters for a starting date in 1282, the year that his patron Andronicus II began his reign.

Nicephorus Gregoras (ca. 1290–ca. 1360) was a polymath of remarkable versatility, as evidenced by his writings on history, theology, philosophy, astronomy, geography, and acoustics. His research in astronomy led him to propose a revision of the calendar, a scheme that anticipated the calendar reform carried out in 1582 by Pope Gregory XIII.

The astronomical work of Theodore Metochites and Nicephorus Gregoras was part of a revival of astronomy in Byzantium that had begun at the end of the thirteenth century, when Greek translations of Arabic astronomical treatises and tables became available in Constantinople. One of the first of such translations was done by Gregory Choniades, who learned Arabic science in Tabriz and founded a school of astronomy in Trebizond circa 1300. One of his students was Manuel of Trebizond, who taught at his master's school. Manuel's most distinguished student was George Chrysokokkes, who wrote on medicine and geography as well as on astronomy.

By the mid-fourteenth century new "Persian Tables" were substituted for older astronomical tables, as in the *Three Books on Astronomy*, a compendium of astronomy published in around 1352 by Theodore Meliteniotes, who became head of the Patriarchal School in Constantinople.

Scientific ideas were also coming to Byzantium from the West, as evidenced by a treatise on the astrolabe translated into Greek in Constantinople circa 1309 from a Latin version of the Arabic original. Another astronomical work, the *Six Wings*, an anonymous thirteenth-century treatise in Hebrew, was translated into Greek in Constantinople in the second quarter of the fifteenth century, apparently having made its way from southern France through Venice to Byzantium.

The fall of Thessalonica to the Ottomans in 1430 led the emperor John VIII (r. 1425–48) to seek help from the West, and he proposed to Pope Martin V that a council be called to help reconcile the Greek and

Latin churches. This eventually gave rise to a council that was convened by Pope Eugenius IV in 1438 at Ferrara, moving to Florence the following year. The council ended on 6 July 1439, when a Decree of Union between the Greek and Latin churches was read in Latin and Greek in the cathedral of Florence, Santa Maria del Fiore, in the presence of the emperor John VIII. But most of the people and clergy of Byzantium rejected the Union, dividing the empire in what were to be the last years of its existence.

The delegates to the Council of Ferrara-Florence included four scholars who were leading figures in the cultural interchange between Byzantium and Italy, one of them representing the Latin Catholic Church and the others the Greek Orthodox faith. The Latin delegate was Nicholas of Cusa; the Greeks were George Gemistus Plethon, George Trapezuntios, and Bessarion of Trebizond.

Nicholas of Cusa (1401–1464) was born at Cues, a village on the Moselle. He was educated at the universities of Heidelberg and Padua, where he received a doctorate in 1423. He entered the priesthood circa 1430 and in 1448 Pope Nicholas V made him a full cardinal, with the titular see of St. Peter in Vincoli in Rome.

Cusa's most important work is his *De Docta Ignorantia* (On Learned Ignorance), completed in 1440. Here he uses mathematics and experimental science in his attempts to determine the limits of human knowledge, particularly in the inability of the human mind to conceive the absolute, which to him was the same as mathematical infinity. He concluded that the universe is infinite in extent, making the idea of a center or of a periphery meaningless. Thus the earth cannot be the center of the universe, and since motion is relative and natural to all bodies, the earth cannot be at rest. A marginal note made by Cusa in one of his manuscripts suggests that the earth cannot be fixed, but rotates on its axis once in a day and a night.

Cusa speculated that the earth might not be the only body on which there were living creatures, and that there might be another earth at the center of the sun's luminous envelope. These and other revolutionary theories led his political rivals to accuse him of pantheism, a charge that he defended himself against in his *Apologia Doctae Ignorantia* (1440), in which he quoted patristic writings and Christian Neoplatonists as the sources of his ideas.

Ten years later Cusa wrote a work entitled *Idiota*, a series of dialogues in which a rhetor, or school man, who represents book learning converses with a layman (Idiota) who stresses the importance of quantitative experimental research. According to Idiota, wisdom clamors in the streets and one can find it in the marketplace, where one sees money being counted, merchandise being weighed, oil being measured, and where one can watch human reason performing its most fundamental function: measurement.

George Gemistus Plethon (ca. 1355–1452), whom Sir Steven Runciman called "the most original of all Byzantine thinkers," was educated in Constantinople and taught there until about 1392. He then went to Mistra in the Peloponnesus, which at the time was ruled by the despot Theodore Palaeologus, second son of the emperor Manuel II (r. 1391–1425). Plethon taught there for the rest of his days, except for a year that he spent as a member of the Byzantine delegation at the Council of Ferrara-Florence. His teaching was dominated by his rejection of Aristotle and his devotion to Plato, who inspired his goal of reforming the Greek world along Platonic lines. His religious beliefs were more pagan than Christian, as evidenced by his treatise *On the Laws*, in which he usually refers to God as Zeus and writes of the Trinity as consisting of the Creator, the World-Mind, and the World-Soul. George Trapezuntios writes of a conversation he had at Florence with Plethon, who told him that the whole world would soon adopt a new religion. When asked if the new religion would be Christian or Muhammadan, Plethon replied, "Neither, it will not be different from paganism."

While the council was deliberating in Florence Plethon delivered a lecture at the palace of Duke Cosimo de' Medici, the subject being the philosophical and religious differences between Platonism and Aristotelianism, in which he eulogized Plato. Cosimo was so inspired by Plethon that he founded a Platonic academy in Florence, which became the center of Renaissance Platonism. Plethon's writings also inspired Marsilio Ficino (1433–99), the first of the great Renaissance Platonists.

During his stay in Florence, Plethon recommended the study of Strabo's *Geography* as a supplement to that of Ptolemy. One of those to whom he spoke of this may have been Paolo Dal Pozzo Toscanelli, whom he met in Florence. Toscanelli later passed the suggestion on to Christopher Columbus, who would have read Strabo in Guarino's Latin translation. Columbus, according to the biography written by his son, was directly influenced by two passages of Strabo's. These were quotes from Eratosthenes and Poseidonus, the first being "If the immensity of the Atlantic did not prevent us, we could sail from Iberia to India along the same parallel," and the second, "If you sail from the west using the east wind you will reach India at a distance of 70,000 stades."

George Trapezuntios (1395–1486) was born on Crete to a family who had moved there from Trebizond; hence his last name. He was a prolific translator from Greek into Latin, which he studied under Guarino in Venice in 1417–18. He taught in Venice, Vicenza, and Mantua before going to Rome, where he served in the papal bureaucracy under Eugenius IV (r. 1431–47) and then taught under Nicholas V (r. 1447–55). He severely criticized Plethon for his attack on Aristotelianism, portraying himself as the champion of medieval scholasticism against its humanist critics, and singling out for special praise Albertus Magnus and Thomas Aquinas, which brought him into conflict with Bessarion, among others.

Bessarion (1403–1472) was born into a family of manual laborers at Trebizond. The metropolitan of Trebizond noticed the boy's intelligence and sent him to school in Constantinople. While at the university he met the Italian humanist Francesco Filelfo, who had been inspired to study in Constantinople after attending the classes of Manuel Chrysoloras in Italy.

At the age of twenty Bessarion became a monk and spent some years at a monastery near Mistra, where he studied under Plethon. He then returned to Constantinople and won renown as a professor of philosophy. He was chosen as one of the delegates to the Council of Ferrara-Florence, and was appointed metropolitan of Nicaea so that he would have appropriate status at the conclave. When the agreement of union was formally proclaimed on 6 July 1439 in the cathedral in Florence, it was first read in Latin by Cardinal Cesarini and then in Greek by Bessarion.

Bessarion's stay in Italy convinced him that Byzantium could survive only in alliance with the West, not just politically, but also by sharing in the cultural life of Renaissance Italy. Disheartened by opposition to the union in Constantinople, he returned to Florence at the end of 1440, by which time he had already been made a cardinal in the Roman Catholic

Church. He spent some time traveling on papal diplomatic missions and served as governor of Bologna from 1450 to 1455, but otherwise from 1443 on he resided in Rome. He was nearly elected pope in 1455, but he lost out when his enemies warned of the dangers of choosing a Greek, and so the cardinals turned to the Catalan Alfonso de Borgia, who was elevated as Callisto III.

Much of Bessarion's energy was spent trying to raise military support in Europe to defend Byzantium against the Turks, but his efforts came to naught since the Ottomans captured Constantinople in 1453 and then took his native Trebizond in 1461, ending the long history of the Byzantine Empire. Thenceforth Bessarion sought to find support for a crusade against the Turks, but to no avail.

Bessarion devoted much of his time to perpetuating the heritage of Byzantine culture by adding to his collection of ancient Greek manuscripts, which he bequeathed to Venice, where they are still preserved in the Marciana Library. The group of scholars who gathered around Bessarion in Rome included George Trapezuntios, whom Bessarion commissioned to translate Ptolemy's *Almagest* from Greek into Latin. Then in 1459 Trapezuntios published an attack on Platonism, suggesting that it led to heresy and immorality. Bessarion was outraged and wrote a defense of Platonism, published in both Greek and Latin. His aim was not only to defend Platonism against the charges made by Trapezuntios, but to show that Plato's teachings were closer to Christian doctrine than those of Aristotle. His book was favorably received, for it was the first general introduction to Plato's thought, which at the time was unknown to most Latins; earlier scholarly works on Platonism had not reached a wide audience.

In 1460 one of Bessarion's diplomatic missions took him to Vienna, where the university had become a center of astronomical and mathematical studies through the work of John of Gmunden (d. 1442), Georg Peurbach (1423–61) and Johannes Regiomontanus (1436–76). John had built astronomical instruments and acquired a large collection of manuscripts, all of which he had bequeathed to the university, thus laying the foundations for the work of Peurbach and Regiomontanus.

Peurbach was an Austrian scholar who had received a bachelor's degree at Vienna in 1448 and a master's in 1453; in the interim he had traveled in France, Germany, Hungary, and Italy. He had served as court astrologer to Ladislaus V, king of Hungary, and then to the king's uncle, the emperor Frederick III. His writings included textbooks on arithmetic, trigonometry, and astronomy, his best-known works being his *Theoricae Novae Planetarum* (New Theories of the Planets) and his *Tables of Eclipses.*

Regiomontanus, originally known as Johann Müller, took his last name from the Latin for his native Königsberg in Franconia. He studied first at the University of Leipzig, from 1447 to 1450, and then at the University of Vienna, where he received his bachelor's degree in 1452, when he was only fifteen, and his master's in 1457. He became Peurbach's associate in a research program that included a systematic study of the planets as well as observations of astronomical phenomena such as eclipses and comets.

Bessarion was dissatisfied with the translation of Ptolemy's *Almagest* that had been done by George Trapezuntios, and he asked Peurbach and Regiomontanus to write an abridged version. They agreed to do so, for Peurbach had already begun work on a compendium of the *Almagest,* but it was unfinished when he died in April 1461. Regiomontantus completed the compendium about a year later in Italy, where he had gone with Bessarion. He spent part of the next four years in the cardinal's entourage and the rest in his own travels, learning Greek and searching for manuscripts by Ptolemy and other ancient astronomers and mathematicians.

Regiomontanus left Italy in 1467 for Hungary, where he served for four years in the court of King Matthias Corvinus, continuing his researches in astronomy and mathematics. He then spent four years in Nuremberg, where he set up his own observatory and printing press. One of the works he printed before his premature death in 1476 was Peurbach's *Theoricae Novae Planetarum;* it would be reprinted in nearly sixty editions until the seventeenth century. He also published his own *Ephemerides,* the first planetary tables ever printed, giving the positions of the heavenly bodies for every day from 1475 to 1506. Columbus is said to have taken the *Ephemerides* with him on his fourth and last voyage to the New World, and to have used its prediction of the lunar eclipse of 29 February 1504 to frighten the hostile natives of Jamaica into submission.

Regiomontanus's most important mathematical work is his *De Triangulis Omnimodis,* a systematic method for analyzing triangles that, together with his *Tabulae Directionum,* marked what a modern historian of mathematics has called "the rebirth of trigonometry."

The astronomical writings of Regiomontanus include the completion of Peurbach's *Epitome of Ptolemy's* Almagest, which he dedicated to Bessarion, a book noted for its emphasis on mathematical methods omitted in other works of elementary astronomy. Copernicus read the *Epitome* when he was a student in Bologna, and at least two propositions in it influenced him in the formulation of his own planetary theory. These propositions seem to have originated with the fifteenth-century Arabic astronomer Ali Qushji and may have been transmitted to Regiomontanus by Bessarion. If so, this would place Bessarion and Regiomontanus in the long chain that leads from Aristarchus of Samos to Copernicus through the Arabic and Latin scholars of the Middle Ages and to the dawn of the Renaissance.

13

THE REVOLUTION OF THE
HEAVENLY SPHERES

Shortly before Christmas 1499 Pope Alexander VI proclaimed that the following year would be a Jubilee, or Holy Year, in which a special indulgence would be granted to all pilgrims who came to Rome and visited the four principal churches of the city, beginning with St. Peter's, whose doors would be open night and day throughout that time. The celebrations went on throughout the year, and on Easter Sunday an estimated 200,000 pilgrims thronged St. Peter's Square for the pope's blessing. The pious monk Petrus Delphinus was led to exclaim "God be praised, who has brought hither so many witnesses to the faith." And Sigismondo de' Conti, the papal secretary, noted in his chronicle, "All the world is in Rome."

Among the pilgrims was the young Nicholas Copernicus, who came to Rome around Easter and remained for a full year, giving lectures on astronomy and mathematics. The Italian Renaissance was in full bloom and Copernicus was in Rome at the height of its glory, before returning to his home in what he called "this remote corner of the earth," in what was then Prussia but is now northern Poland.

Copernicus was born on 19 February 1473 at Torun, a town on the Vistula 110 miles northwest of Warsaw. His name was originally Niklas Koppernigk, which he Latinized as Nicholas Copernicus after he went to college. He was the youngest of four children of a prosperous merchant, the others being his brother Andreas and his sisters Barbara and

Katherina. Their father died in 1483, whereupon the children were adopted by their maternal uncle Lucas Watzenrode, a priest who had studied at the universities of Cracow, Cologne, and Bologna. In 1489 Lucas become bishop of Ermland, also known as Varmia, one of the four provinces into which Prussia was then divided. The bishop's palace was at Heilsberg (Lidzbark Warminski), 140 miles northeast of Torun, while the cathedral of the bishopric was at Frauenburg (Frombork), on the shore of the long lagoon east of Danzig (Gdansk). Nicholas and Andreas stayed with their uncle Lucas at his palace in Heilsberg, while Barbara entered a convent and Maria married a merchant in Cracow.

In the autumn of 1491 Nicholas and Andreas were sent by their uncle Lucas to the University of Cracow, where they enrolled in the faculty of arts. They remained there for three or four years, but left without earning a degree. During that time Nicholas is known to have taken courses in mathematics, astronomy, astrology, and geography, and his reading included Cicero, Virgil, Ovid, and Seneca.

The renowned Polish astronomer Albert Brudzewski was lecturing at the University of Cracow at the time and Nicholas would have read his works, although there is no record of their having met. Brudzewski had published a commentary on the planetary theory of Peurbach, in which he put forward his own theory that the celestial orbs are not spheres but circles. Brudzewski also used a mathematical method analogous to one employed by the Arabic astronomers Nasr al-Din al-Tusi and Ibn al-Shatir, similar to a model that Copernicus would later use in his heliocentric theory.

The textbooks Copernicus read in his courses in mathematics, astronomy, and astrology included works by Euclid, Ptolemy, Sacrobosco, Peurbach, and Regiomontanus. The works of a number of Arabic astronomers were available in Cracow at that time, including those of Masha'allah, al-Farghani, al-Kindi, Thabit ibn Qurra, and Jabir ibn Aflah. Copernicus also bought a number of books in Johann Haller's bookshop in Cracow, including the *Alphonsine Tables* and the *Tabulae Directionum* of Regiomontanus, which he had bound together with parts of Peurbach's *Tables of Eclipses* and tables of planetary latitudes.

Nicholas and Andreas left Cracow early in 1496 to live with their uncle Lucas in the bishop's palace at Heilsberg. Lucas nominated

Nicholas and Andreas to be canons of Frauenburg Cathedral, but at first his efforts were unsuccessful. Nicholas was finally made a canon in 1497, and Andreas was elected in 1499. Both of them were elected in absentia, for in the fall of 1496 Nicholas had gone off to study at the University of Bologna; Andreas joined him there two years later. They were both in the faculty of law and enrolled in the Natio Germanorum, the largest of the "nations" into which foreign students were organized at Bologna.

The brothers seem to have stayed as paying guests in the house of Domenico Maria da Novara (1454–1504) of Ferrara, a professor of astronomy at the university. Nicholas believed that he was "not so much the pupil as the assistant and witness of observations of the learned Dominicus Maria," as his friend Rheticus later relayed, quoting Copernicus. One of the observations at Bologna concerned a lunar occultation of the star Aldabaran, which Copernicus says they made "after sunset on the seventh day before the Ides of March, in the year of Christ 1497."

Copernicus took his M.A. degree in law at Bologna in 1499, after which he went to Rome for a year. According to Rheticus, while in Rome Copernicus "lectured on mathematics before a large audience of students and a throng of great men and experts in this branch of knowledge." Copernicus observed a lunar eclipse in Rome on 6 November 1500. He compared it to an eclipse observed by Ptolemy at Alexandria in the "nineteenth year of Hadrian" (A.D. 136–137), his purpose being "to determine the positions of the moon's movement in relation to the established beginnings of calendar years."

Nicholas and Andreas returned to Poland in May 1501. On 27 July of that year they made an appeal to the authorities of their chapter in Frauenburg, asking for a two-year extension of their leave so that they could complete their studies in Italy. The chapter accepted and in August they left Frauenburg for Italy, Andreas to complete his degree in canon law in Bologna and Nicholas to study medicine in Padua.

Nicholas enrolled at the University of Padua in the fall of 1501, studying law as well as medicine. He interrupted his studies at Padua after two years to enroll at the University of Ferrara, where on 31 May 1503 he received the degree of doctor of canon law. He then returned to Padua to continue his study of medicine, but he left in 1505 without obtaining a degree.

After his return from Italy Copernicus joined his uncle Lucas at

Heilsberg Castle, the official residence of the bishops of Ermland. During the next six years Copernicus served as private physician and secretary of state to his uncle. He also represented Ermland in the Polish diets, or parliaments, and served on diplomatic missions to the King of Poland in Cracow. Early in 1512 he accompanied his uncle to Cracow to attend the wedding of King Sigismund and the coronation of his bride. But on the way home Bishop Lucas died in Torun, on 29 March 1512, after which his body was brought back to Frauenburg and buried in the cathedral.

After his uncle's death Copernicus left Heilsberg and returned to Frauenburg, where he resumed his duties as a canon. His apartment there was at the northeast corner of the enclosure wall of the cathedral, in a turret that is still called the Tower of Copernicus, which also served as his observatory. The first observation he recorded there was on 5 June 1512, when he noted that Mars was in opposition—that is, the planet rose at sunset and set at sunrise, since it was diametrically opposite the sun in the celestial sphere. This was the first of at least twenty-five observations Copernicus would make at Frauenburg, where he also developed the mathematical methods that he used in his new astronomical theory.

Around this time Copernicus began writing a work entitled *Nicolai Copernici de Hypothesibus Motuum Caelestium a Se Constitutis Commentariolus* (Nicholas Copernicus, Sketch of His Hypotheses for the Celestial Motions). This came to be known as the *Commentariolus*, or Little Commentary, the first notice of the new astronomical theory Copernicus had been developing. He gave written copies of this short treatise to a few friends but never published it in book form. Only two manuscript copies have survived, one of which was first published in Vienna in 1878. The earliest record of the *Commentariolus* is a note made in May 1514 by a Cracow professor, Matthias de Miechow, who writes that he had in his library "a manuscript of six leaves expounding the theory of an author who asserts that the earth moves while the sun stands still." Matthias was unable to identify the author of this treatise, since Copernicus, with his customary caution, had not written his name on the manuscript, for, as he later admitted, he feared that he would be censured and ridiculed for his revolutionary ideas. But there is no doubt that the manuscript was by Copernicus, because the author made a marginal note that he reduced all his calculations "to the meridian of Cracow, because . . .

Fromberk [Frauenburg] . . . where I made most of my observations . . . is on this meridian as I infer from lunar and solar eclipses observed at the same time in both places."

The introduction to the *Commentariolus* discusses the theories of Greek astronomers concerning "the apparent motion of the planets," noting that the homocentric spheres of Eudoxus were "unable to account for all the planetary motions" and were supplanted by Ptolemy's "eccentrics and epicycles, a system which most scholars finally accepted." But Copernicus took exception to Ptolemy's use of the equant, which led him to think of formulating his own planetary theory, "in which everything would move uniformly about its proper center, as the rule of absolute motion requires."

Copernicus goes on to say that after setting out to solve "this very difficult and almost insoluble problem," he finally arrived at a solution that involved "fewer and much simpler constructions than were formerly used," provided that he could make certain assumptions, seven in number.

The assumptions are: that there is not a single center for all the celestial circles, or spheres; that the earth is not the center of the universe, but only of its own gravity and of the lunar sphere; that the sun is the center of all the planetary spheres and of the universe; that the earth's radius is negligible compared to its distance from the sun, which in turn is "imperceptible in comparison to the height of the firmament"; that the apparent diurnal motion of the stellar sphere is due to the rotation of the earth on its axis; that the daily motion of the sun is due to the combined effect of the earth's rotation and its revolution around the sun; and that "the apparent retrograde and direct motion of the planets arise not from their motion but from the earth's." He then concludes that "the motion of the earth alone, therefore, suffices to explain so many inequalities in the heavens."

Copernicus goes on to describe the "Order of the Spheres" in his heliocentric system, in which the time taken by a planetary sphere to make one revolution increases with the radius of its orbit.

The celestial spheres are arranged in the following order. The highest is the immovable sphere of the fixed stars, which contains and gives position to all things. Beneath it is Saturn, which Jupiter follows, then Mars. Below Mars is the sphere on which we revolve, then Venus; last is Mercury. The lunar sphere revolves around the center of the earth and moves with the earth like an epicycle. In the same order, also, one planet surpasses another in speed of revolution, accordingly as they trace greater or smaller circles. Thus Saturn completes its revolution in thirty years, Jupiter in twelve, Mars in two and one-half, and the earth in one year; Venus in nine months, Mercury in three.

Copernicus used the same system of epicycles that Ptolemy and all of his successors had employed in the geocentric model. He concludes the *Commentariolus* by summarizing the number of circles (i.e., deferents, or primary circles) and epicycles (secondary loops) required to describe all of the celestial motions in his heliocentric system: "Then Mercury runs on seven circles in all; Venus on five; the earth on three, and round it the moon on four; finally Mars, Jupiter and Saturn on five each. Altogether, therefore, thirty-four circles suffice to explain the entire structure of the universe and the entire ballet of the planets."

The first indication that the new theories of Copernicus had reached Rome came in the summer of 1533, when the papal secretary Johann Widmanstadt gave a lecture entitled "Copernicana de Motuu Terra Sentential Explicani" (An Explanation of Copernicus's Opinion of the Earth's Motion) before Pope Clement VII and a group that included two cardinals and a bishop. After the death of Pope Clement, on 25 September 1534, Widmanstadt entered the service of Cardinal Nicholas Schönberg, who as papal nuncio in Prussia and Poland had undoubtedly heard of Copernicus years before. Schönberg wrote to Copernicus on 1 November 1536—the letter may have been drafted by Widmanstadt— urging Copernicus to publish a book on his new cosmology and to send him a copy.

Despite this encouragement Copernicus made no move to publish his researches, but his attitude changed in the spring of 1539, when he received an unexpected visit from a young German scholar, Georg Joachim van Lauchen, who called himself Rheticus (1514–1574). Rheticus, who although only twenty-five was already a professor of mathematics at the Protestant University of Wittenberg, explained that he was deeply interested in Copernicus's new cosmology; Copernicus received him hospitably and permitted him to study the manuscript he had written to explain his theories. During the next ten weeks Rheticus worked with Copernicus in studying the manuscript, which he then summarized in a treatise entitled *Narratio Prima* (First Narrative), intended as an introduction to the Copernican theory. This was written in the form of a letter from Rheticus to his friend Johann Schöner, under whom he had studied at Wittenberg. The *Narratio Prima* was published at Danzig in 1540 with the approval of Copernicus, who is referred to by Rheticus as "my teacher" in the introductory section where he describes the scope of the Copernican cosmology.

Rheticus then goes into each of the books in detail, adding an astrological prediction of his own after his account of the Copernican theory in "The Eccentricity of the Sun and the Motion of the Solar Apogee." Rheticus believed that world history followed the same cycle as the eccentricity of the sun's orbit (observed from the earth), and that the completion of its next cycle would coincide with the downfall of the Muhammadan faith, following which, he said, "We look forward to the coming of our Lord Jesus Christ when the center of the eccentric reaches the other boundary of mean value, for it was in that position at the creation of the world."

Rheticus does not mention the heliocentric theory until after the section entitled "General Considerations Regarding the Motions of the Moon, Together with the New Lunar Hypotheses." There he says that the new model explains the retrograde motion of the planets "by having the sun occupy the center of the universe, while the earth revolves instead of the sun on the eccentric."

The *Narratio Prima* proved to be so popular that a second edition was printed at Basel the following year. But Copernicus still hesitated to publish his manuscript, which he sent for safekeeping to his old friend Tiedemann Giese, the bishop of Culm. Finally, in the autumn of 1541, Giese received permission from Copernicus to send his manuscript to Rheticus, who was to take it to the press of Johannes Petreius in Nuremberg for publication. The title chosen for the book was *De Revolutionibus Orbium Coelestium Libri VI* (Six Books Concerning the Revolutions of the Heavenly Spheres). The title stems from the fact that Copernicus believed the celestial bodies to be embedded in the same crystalline spheres—or, rather, spherical shells—as those first proposed by Aristotle, though he had them revolving around the sun rather than the earth.

Toward the end of the following year Copernicus suffered a series of strokes that left him half paralyzed, and it was obvious to his friends that his end was near. Tiedemann Giese wrote on 8 December 1542 to George Donner, one of the canons at Frauenburg, asking him to look after Copernicus in his last illness. "I know that he always counted you among his truest friends. I pray therefore, that if his occasions require, you will stand by him and take care of the man whom you, with me, have ever loved, so that he may not lack brotherly help in his distress, and that we may not appear ungrateful to a friend who has richly deserved our love and gratitude."

Meanwhile, Rheticus had taken a leave of absence from the University of Wittenberg in May 1542 to supervise the printing of *De Revolutionibus* in Nuremberg. Five months later he left Nuremberg to take up a post at the University of Leipzig, leaving responsibility for the book in the hands of Andreas Osiander, a local Lutheran clergyman. Osiander took it upon himself to add an anonymous introduction entitled "Ad Lectorem" (To the Reader), which was to be the cause of considerable controversy regarding the Copernican theory.

De Revolutionibus finally came off the press in the spring of 1543, with the publisher's blurb, probably also written by Osiander, printed directly below the title. "You have in this recent work, studious reader, the motion of both the fixed stars and the planets recovered from ancient as well as recent observations and outfitted with wonderful new and admirable hypotheses. You also have most expeditious tables from which you can easily compute the positions of the planets for any time. Therefore buy, read, profit."

The first printed copy of *De Revolutionibus* was sent to Copernicus, and according to tradition it reached him a few hours before he died, on 24 May 1543. Tiedemann Giese described the last days of Copernicus in a letter to Rheticus: "He had lost his memory and mental vigor many days before; and he saw his completed work only at his last breath upon the day that he died."

The introduction to *De Revolutionibus,* the "Ad Lectorum" written by Osiander, is addressed "To the Reader Concerning the Hypotheses of This Work." It says that the book is designed as a mathematical device

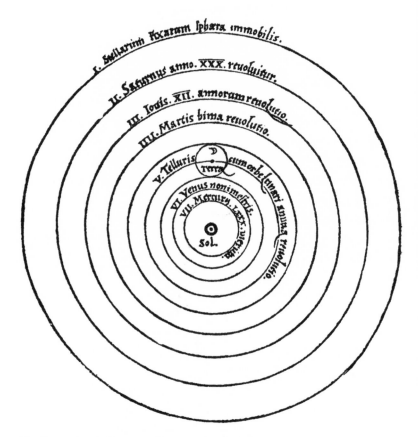

The Copernican heliocentric theory, *De Revolutionibus*, 1543.

for calculation and not as a real description of nature. The "Ad Lectorum" was intended to deflect criticism of the heliocentric cosmology by those who thought that it contradicted the Bible, particularly the passage in the Book of Joshua that says, "The sun stood still in the middle of the sky and delayed its setting for almost a whole day." Martin Luther, referring to the Copernican theory, had already been quoted as remarking, "People give ear to an upstart astrologer who strove to show that the earth revolves, not the heavens, or the firmament, the sun and the moon. This fool wishes to reverse the entire science of astronomy, but sacred Scripture tells us that Joshua commanded the Sun to stand still and not the Earth." Copernicus himself had been worried about such criticism, as evidenced by his statement in the preface of *De Revolutionibus*, which he dedicated to Pope Paul III: "I can reckon easily enough, Most Holy Father, that as soon as certain people learn that in these books of mine which I have written about the revolutions of the spheres of the world I attribute certain motions to the terrestrial globe they will immediately shout to have me and my opinion hooted off the stage."

The first eight chapters of Book I of *De Revolutionibus* give a greatly simplified description of the Copernican cosmology and its philosophical basis. Copernicus begins with arguments for the spherical nature of the universe; the sphericity of the earth, moon, sun, and planets; and the uniform circular motion of the planets around the sun. He shows how the rotation of the earth on its axis, together with its revolution about the sun, can easily explain the observed motions of the celestial bodies. The absence of stellar parallax he explains by the fact that the radius of the earth's orbit is negligible compared to the distance of the fixed stars. All of the arguments on physical grounds against the earth's motion are then refuted, using in most cases the explanations given by Nicholas of Cusa.

Chapter 9 is entitled "Whether Many Movements Can Be Attributed to the Earth, and Concerning the Center of the World." Here Copernicus abandons the Aristotelian doctrine that the earth is the sole source of gravity and instead takes the first step toward the Newtonian theory of universal gravitation, writing, "I myself think that gravity or heaviness is nothing except a certain natural appetency implanted in the parts by the divine providence of the universal Artisan, in order that they should unite with one another in their oneness and wholeness and come together in the form of a globe."

Chapter 10 is entitled "On the Order of the Celestial Orbital Circles." Here Copernicus removes the ambiguity concerning Mercury and Venus, which in the Ptolemaic model were sometimes placed "above" the sun and sometimes "below." The Copernican system has Mercury as the closest planet to the sun, followed by Venus, Earth, Mars, Jupiter, and Saturn, surrounded by the sphere of the fixed stars, and with the moon orbiting the earth. This model is simpler and more harmonious than Ptolemy's, for all of the planets revolve in the same sense, with velocities decreasing with their distance from the sun, which sits enthroned at the center of the cosmos.

> In the center of all the celestial bodies rests the Sun. For who would place this lamp of a very beautiful temple in another or better place than this wherefrom it can illuminate everything at the same time. As a matter of fact, not unhappily do some call it the lantern, others the mind and still others, the pilot of the world.... And so the sun, as if resting on a kingly throne, governs the family of stars which wheel around.

Chapter 11 is "A Demonstration of the Threefold Movement of the Earth," while the remaining three chapters of Book I are concerned with the application of plane and spherical geometry and trigonometry to problems in astronomy. The three motions to which Copernicus refers are the earth's rotation on its axis, its revolution around the sun, and a third conical motion, which he introduced to keep the earth's axis pointing in the same direction while the crystalline sphere in which it was embedded rotated annually. The period of this supposed third motion he took to be slightly different from the time it takes the earth to rotate around the sun, the difference being due to the very slow precession of the equinoxes.

Book II is a detailed introduction to astronomy and spherical trigonometry, together with mathematical tables and a catalog of the celestial coordinates of 1,024 stars, most of them derived from Ptolemy, adjusted for the precession of the equinoxes.

Book III is concerned with the precession of the equinoxes and the movement of the earth around the sun. Here the theory is unnecessarily complicated, since Copernicus, besides combining precession with his "third motion" of the earth, inherited two effects from his predecessors, one of them spurious. The first effect was the mistaken notion, stemming from the trepidation theory, that the precession was not constant but variable; the other was the variation in the inclination of the ecliptic.

Book IV deals with the motion of the moon around the earth; Books V and VI study the motions of the planets. Here, as with the motions of the sun, Copernicus used eccentrics and epicycles just as Ptolemy had done, though his conviction that the celestial motions were combinations of circular motion at constant angular velocity made him refrain from using the Ptolemaic device of the equant. Because of the complexity of the celestial motions, Copernicus was forced to use about as many circles as had Ptolemy, and so there was little to choose from between the two theories so far as economy was concerned, and both were capable of giving results of comparable accuracy. The advantages of the Copernican system were that it was more harmonious; it removed the ambiguity about the order of the inner planets; it explained the retrograde motion of the planets as well as their variation in brightness; and it allowed both the order and the relative sizes of the planetary orbits to be determined from observation without any additional assumptions.

Copernicus mentions some of the Arabic astronomers whose observations and theories he used in *De Revolutionibus*, namely, al-Battani, al-Bitruji, al-Zarqali, Ibn Rushd (Averroës), and Thabit ibn Qurra. But he does not mention Nasr al-Din al-Tusi or Ibn al-Shatir, although recent research shows that he used a mathematical method that had been developed by them. This is the so-called al-Tusi couple, which Copernicus used in Book III, Chapter 4 of *De Revolutionibus*. There is no definite evidence that Copernicus knew of al-Tusi or al-Shatir, but current opinion is that he was aware of their works, which were known to some of his contemporaries.

Copernicus refers to Aristarchus of Samos thrice in *De Revolutionibus*, twice regarding his predecessor's measurement of the inclination of the ecliptic and once concerning his measurement of the length of the solar year. But nowhere does he mention that Aristarchus had, in the mid-third century B.C., proposed that the sun and not the earth was the center of the cosmos. Copernicus had referred to the heliocentric theory of Aristarchus in his original manuscript, but he'd deleted it from the edition of *De Revolutionibus* printed in 1543. The suppressed passage, which had been in the last paragraph of Book I, Chapter 11, reads:

> Though the courses of the Sun and the Moon can surely be demonstrated on the assumption that the Earth does not move, it does not work so well with the other planets. Probably for this and other reasons, Philolaus perceived the mobility of the Earth, a view also shared by Aristarchus of Samos, so some say, not impressed by that reasoning which Aristotle cites and refutes.

Copernicus is known to have possessed a copy of George Valla's *Outline of Knowledge*, printed by Aldus Manutius at Venice in 1501, which included a translation of a work by Aetius (Pseudo-Plutarch) containing two references to Aristarchus. One has Aristarchus "assuming that the heavens are at rest while the earth revolves along the ecliptic, simultaneously rotating about its own axis"; the other says that in his theory the earth "spins and turns, which Seleucus afterwards advanced as an established opinion."

Copernicus was almost certainly familiar with Archimedes' *Sand Reckoner*, which contains the earliest reference to the heliocentric theory of Aristarchus. There Archimedes says that Aristarchus explains the lack of stellar parallax in his heliocentric theory by supposing that the radius of the earth's orbit around the sun is negligible compared to the distance of the stars. This is essentially the same explanation given by Copernicus in his *Commentariolus*, where in the fourth of his assumptions he states that "the distance from the earth to the sun is imperceptible in comparison to the height of the firmament." Copernicus uses this same argument in *De Revolutionibus*, where at the end of Book I, Chapter 10, he contrasts the retrograde motion of the planets with the unchanging array of the stars, noting, "How exceedingly fine is the godlike work of the Best and Greatest Artist!"

Thus it would seem that Copernicus was aware of Aristarchus's heliocentric theory and that he chose to suppress mention of it in *De Revolutionibus*, perhaps so as not to lessen the importance of his own life's work, setting the celestial orbs in motion around the sun rather than the earth.

Cosmology would never again be the same after the publication of *De Revolutionibus*. The world picture was now irrevocably changed, an intellectual revolution started by an obscure canon working alone in what he called a "remote corner of the earth," reviving a theory that had first been proposed eighteen centuries before by an almost forgotten Greek astronomer.

14

THE DEBATE OVER THE TWO
WORLD SYSTEMS

Copernicus received many posthumous accolades after the publication of *De Revolutionibus*, but the praise was primarily for his success in giving a mathematical description of the motion of the celestial bodies rather than for his sun-centered cosmology.

The first astronomical tables based on the Copernican theory were produced by Erasmus Reinhold (1511–53), a professor of mathematics at the University of Wittenberg at the same time as Rheticus. When Rheticus returned to Wittenberg in September 1541 with the manuscript of *De Revolutionibus*, he probably showed it to Reinhold. The following year Reinhold published his commentary on Peurbach's *Theoricae Novae Planetarum*, where he wrote of Copernicus in the preface, "I know of a modern scientist who is exceptionally skillful. He has raised a lively expectancy in everybody. One hopes that he will restore astronomy." Reinhold set out to produce a more extensive version of the planetary tables in *De Revolutionibus*. These were published in 1551 as *The Prutenic Tables*; in the introduction he praises Copernicus but is silent about his heliocentric theory. Reinhold also wrote a commentary on *De Revolutionibus*, but it was unfinished when he died of the plague in 1553.

The Prutenic Tables was the first complete compilation of planetary tables prepared in Europe for three centuries. They were demonstrably superior to the older tables, which were now out of date, and so they were used by most astronomers, lending legitimacy to the Copernican theory even when those who used them did not acknowledge the sun-centered cosmology of Copernicus. As the English astronomer Thomas Blundeville wrote in the preface to an astronomy text in 1594: "Copernicus . . . affirmeth that the earth turneth about and that the sun standeth still in the midst of the heavens, by help of which false supposition he hath made truer demonstrations of the motions and revolutions of the celestial spheres, than ever were made before."

The English mathematician Robert Recorde (1510–58) was one of the first to lend some support to the Copernican theory. He discusses the theory in his *Castle of Knowledge* (1551), written in the form of a dialogue between a master and a scholar concerning Ptolemy's arguments against the earth's motion. After the scholar sums up these arguments, the master nevertheless presents the Copernican theory in a very positive manner.

> That is trulye to be gathered: howe be it, COPERNICUS, a man of greate learning, of much experience, and of wonderfull diligence in observation, hath renewed the opinion of ARISTARCHUS SAMIUS, and affirmith that the earthe not only moveth circularlye about his own centre, but also may be, yea and is, continually out of the precise centre of the world 38 hundred thousand miles.

Five years later the *Ephemeris,* a set of astronomical tables, was printed in London for the year 1557 by John Feild, based on the work of Copernicus and Reinhold. The introduction was written by John Dee (1527–1608), Queen Elizabeth's astrologer, who noted that he had persuaded Feild to compile the *Ephemeris,* based on the Copernican theory, since the older tables were no longer satisfactory. Dee praised Copernicus for his "more than Herculean" effort in restoring astronomy, though he said that this was not the place to discuss the heliocentric theory itself.

A second edition of *De Revolutionibus* was published at Basel in 1566, and copies of it made their way to Italy and England. The English astronomer Thomas Digges (ca. 1546–95), a pupil of Dee's, obtained a copy of *De Revolutionibus;* it has survived in the library of Geneva University, along with a note he wrote on the title page, *"Vulgi opinio Error"* ("The common opinion errs"), indicating that he was one of the few sixteenth-century scholars who accepted the Copernican theory.

Digges did a free English translation of Chapters 9 through 11 of the first book of *De Revolutionibus*, adding it to his father's perpetual almanac, *A Prognostication Everlasting*, and publishing them together in 1576 as *Perfit Description of the Caelestiall Orbes, according to the most ancient doctrines of the Pythagoreans lately revived by Copernicus and by Geometricall Demonstrations approved*. Digges stated that he had included this excerpt from *De Revolutionibus* in the almanac "so that Englishmen might not be deprived of so noble a theory."

The book was accompanied by a large folded map of the sun-centered universe in which the stars were not confined to the outermost celestial sphere but scattered outward indefinitely in all directions. Digges thus burst the bounds of the medieval cosmos, which till then had been limited by the ninth celestial sphere, the one containing the supposedly fixed stars, which in his model extended to infinity.

The geocentric celestial spheres were still very much a part of the general worldview in late-sixteenth-century England, as is evident from Christopher Marlowe's play *The Tragical History of Dr. Faustus*. As soon as Faustus has made his pact with the devil he begins to question Mephistophilis about matters beyond the ken of mortals, beginning with the celestial spheres.

> *Speak, are there many spheres above the moon?*
> *Are all celestial bodies but one globe,*
> *As is the substance of this centric earth?*

The concept of an infinite universe was one of the revolutionary ideas for which the Italian mystic Giordano Bruno (1548–1600) was condemned by the Catholic Church, which had him burned at the stake in Rome on 17 February 1600. At the beginning of his dialogue in *The Infinite Universe and the Worlds*, published in 1584, Bruno says, through one of his characters, that in this limitless space there are innumerable worlds similar to our earth, each of them revolving around its own star-sun. "There are then innumerable suns, and an infinite number of earths revolve around these suns, just as the seven [the five visible planets plus the earth and its moon] we can observe revolve around this sun which is close to us."

Bruno's universe was not only infinite but dynamic, in contrast to the

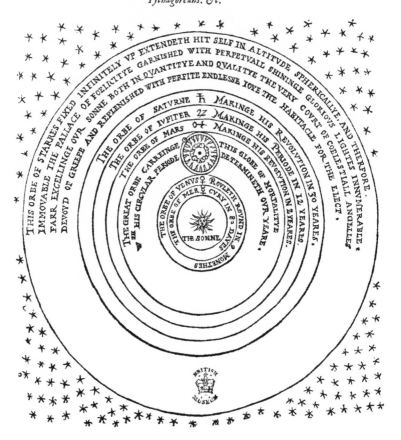

A model of the Copernican system from Thomas Digges's *A Prognostication Everlasting*, 1576.

finite cosmos of Aristotle, for whom the celestial region was immutable. Here he took his inspiration from the atomic theory of Democritus as expressed by Lucretius in his *De Rerum Natura*, which had been rediscovered in 1417. According to Bruno, the living universe is limited neither in its extent nor in its constantly changing multiplicity:

> There are no ends, boundaries, limits or walls which can defraud or deprive us of the infinite multitude of things....

Thus Democritus and Epicurus, who maintained that everything throughout infinity suffered renewal and restoration, understood these matters more truly than those who at all costs maintain a belief in the immutability of the Universe, alleging a constant and unchanging number of particles of identical material that perpetually undergo transformation, one into another.

The concept of an infinite universe appears also in the work of the English scientist William Gilbert (1544–1603), who may have been influenced by Thomas Digges and Giordano Bruno. Gilbert's *De Magnete*, published in 1600, was the first work on magnetism since that of Petrus Peregrinus in the thirteenth century. The sixth and final book of this work was devoted to Gilbert's cosmological theories, in which he rejected the crystalline celestial spheres of Aristotle and said that the apparent diurnal rotation of the stars was actually due to the axial rotation of the earth, which he believed to be a huge magnet. His rejection of diurnal stellar motion was due to his belief that the stars were limitless in number and extended to infinity, so that it was ridiculous to think that they rotated nightly around the celestial pole.

Meanwhile, astronomy was being revolutionized by the Danish astronomer Tycho Brahe (1546–1601), who in the last quarter of the seventeenth century made systematic observations of significantly greater accuracy than any ever done in the past, all just before the invention of the telescope.

Tycho, born to a noble Danish family, became passionately interested in astronomy while still a young boy and spent his nights observing the heavens. He enrolled at the University of Copenhagen at the age of thirteen and subsequently continued his studies at the universities of Leipzig, Basel, and Rostock. The astronomy books he studied included Sacrobosco's *De sphaera*, which had been in use since the thirteenth century, and other texts based on the homocentric spheres of Aristotle and the epicycles and eccentrics of Ptolemy. He was keenly interested in the new theory of Copernicus, whom he called "a second Ptolemy."

Tycho made his first important observation in August 1563, when he noted a conjunction of Saturn and Jupiter. He found that the *Alfonsine Tables* were a month off in predicting the date of the conjunction, and that the *Prutenic Tables* were several days in error. This convinced Tycho that new tables were needed, and that they should be based on more accurate, precise, and systematic observations, which he would make with instruments of his own design in his own observatory. The first of Tycho's observatories was at Augsburg, Germany, where he lived in the years 1569–71. The instruments that he designed and built for his observatory included a great quadrant with a radius of some nineteen feet for measuring the altitude of celestial bodies. He also constructed a huge sextant with a radius of fourteen feet for measuring angular separations, as well as a celestial globe ten feet in diameter on which to mark the positions of the stars in the celestial map that he began to create.

Tycho returned to Denmark in 1571, and on 11 November of the following year he began observing a nova, or new star, that suddenly appeared in the constellation Cassiopeia, exceeding even the planet Venus in its brilliance. (It is now known that a nova is a star that is exploding at the end of its evolutionary cycle, releasing an enormous amount of energy for a few months.) Tycho's measurements indicated that the nova was well beyond the sphere of Saturn, and the fact that its position did not change showed that it was not a comet. This was clear evidence of a change taking place in the celestial region, where, according to Aristotle's doctrine, everything was perfect and immutable.

The nova eventually began to fade, its color changing from white to yellow and then red, finally disappearing from view in March 1574. By then Tycho had written a brief tract entitled *De Nova Stella* (The New Star), which was published at Copenhagen in May 1573. After presenting the measurements that had led him to conclude that the new star was in the heavens beyond the planetary spheres, Tycho expressed his amazement at what he had observed. "I doubted no longer," he wrote. "In truth, it was the greatest wonder that has ever shown itself in the whole of nature since the beginning of the world, or in any case as great as [when the] Sun was stopped by Joshua's prayers."

The tract impressed King Frederick II of Denmark, who gave Tycho an annuity along with the small offshore island of Hveen, in the Øresund Strait, north of Copenhagen, the revenues of which would enable him to build and equip an observatory. Tycho settled on Hveen in 1576, calling the observatory Uraniborg, meaning "City of the Heavens." The astronomical instruments and other equipment of what came to be a large research center were so numerous that Tycho was forced to build an annex called Stjernborg, "City of the Stars," with subterranean chambers to shield the apparatus and researchers from the elements. That same year Tycho and his assistants began a series of observations of unprecedented accuracy and precision that would continue for the next two decades, laying the foundations for what would prove to be the new astronomy.

A spectacular comet appeared in 1577 and Tycho made detailed observations that led him to conclude that it was farther away than the moon, in fact even beyond the sphere of Mercury, and that it was in orbit around the sun among the outer planets. This contradicted the Aristotelian doctrine that comets were meteorological phenomena occurring below the sphere of the moon. He was thus led to reject Aristotle's concept of the homocentric crystalline spheres, and he concluded that the planets were moving independently through space.

Tycho's star catalog was based on systematic measurements of the coordinates of twenty-one principal stars, with a mean error, compared to modern values, of less than 40 seconds of arc, far less than that of any of his predecessors. Comparing the coordinates of the twenty-one principal stars in his catalog with those measured from antiquity until to his own time, Tycho computed a value for the rate of precession of the equinoxes equal to 51 seconds of arc per year, as compared to the modern value of 50.23 seconds. He correctly assumed the precession to be uniform, making no mention of the erroneous Islamic trepidation theory, which had caused unnecessary problems for Copernicus.

Despite his admiration for Copernicus, Tycho rejected the heliocentric theory, both on physical grounds and on the absence of stellar parallax; in the latter case he did not take into account the argument made by Archimedes and Copernicus that the stars were too far away to show any parallactic shift. Tycho rejected both the diurnal rotation of the earth as well as its annual orbital motion, retaining the Aristotelian belief that the stars rotated nightly around the celestial pole.

Faced with the growing debate between the Copernican and Ptolemaic theories, Tycho was led to propose his own planetary model, in which Mercury and Venus revolved around the sun, which together with the other planets and the moon orbited around the stationary earth. Tycho believed that his model combined the best features of both the Ptolemaic and the Copernican theories, since it kept the earth stationary and explained why Mercury and Venus were never very far from the sun.

Tycho's patron Frederick II died in 1588 and was succeeded by his son Christian IV, who was then eleven years old. When Christian came of age, in 1596, he informed Tycho that he would no longer support his astronomical research. Tycho was thus forced to abandon Uraniborg, taking with him all of his astronomical instruments and records, hoping to find a new royal patron.

Tycho moved first to Copenhagen and then in turn to Rostock and Wandsburg Castle, outside Hamburg. He remained for two years at Wandsburg Castle, where in 1598 he published his *Astronomiae Instauratae Mechanica*, a description of all of his astronomical instruments. He sent copies of his treatise to all of the wealthy and powerful people who might be interested in supporting his further researches. He appended his star catalog to the copy he presented to the emperor Rudolph II, who agreed to support Tycho's work, appointing him as the court astronomer.

Thus in 1600 Tycho moved to Prague, where he set up his instruments and created a new observatory at Benatky Castle, several miles northeast of the city. Soon afterward he hired an assistant named Johannes Kepler (1571–1630), a young German mathematician who had sent him an interesting treatise on astronomy, the *Mysterium Cosmographicum*.

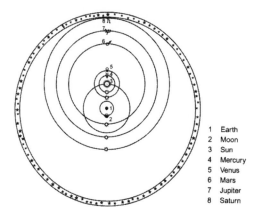

1 Earth
2 Moon
3 Sun
4 Mercury
5 Venus
6 Mars
7 Jupiter
8 Saturn

The Tychonic system, showing Mercury and Venus in orbit around the sun, which orbits the earth along with other planets and the moon, with the stars in the outermost sphere.

Kepler was born on 27 December 1571 in Weil der Stadt in southwestern Germany. His father was an itinerant mercenary soldier, his mother a fortune-teller who at one point was accused of being a witch and almost burned at the stake. The family moved to the nearby town of Lemberg, where Kepler was enrolled in one of the excellent Latin schools founded by the Duke of Württemberg. His youthful interest in astronomy had been stimulated by seeing the comet of 1577 and a lunar eclipse in 1580.

In 1589 Kepler entered the University of Tübingen, where, in addition to his studies in mathematics, physics, and astronomy, he was influenced by Platonism, Pythagoreanism, and the cosmological ideas of Nicholas of Cusa. His mathematics lectures were based on the works of Euclid, Archimedes, and Apollonius of Perge. (As Kepler later said, "How many mathematicians are there, who would toil through the *Conics* of Apollonius of Perge?")

Kepler was particularly influenced by his professor of astronomy, Michael Maestlin, from whom he first learned of the heliocentric theory. In the introduction to his first book, the *Mysterium Cosmographicum*, Kepler wrote of his excitement on discovering the work of Copernicus, which he described as "a still unexhausted treasure of truly divine insight into the magnificent order of the whole world and of all bodies."

Kepler received his master's degree at Tübingen in 1591, after which he studied theology there until 1594, when he was appointed a teacher of mathematics at the Protestant seminary in the Austrian town of Graz. A year after his arrival in Graz, Kepler came up with an idea that he thought explained the arrangement and order of the heliocentric planetary system. He had learned from his reading of Euclid that there are five and only five regular polyhedra, the so-called Platonic solids, in which all of the faces are equal as well as equilateral—the cube, tetrahedron, dodecahedron, icosahedron, and octahedron—and it occurred to him that they were related to the orbits of the earth and the five other planets. He explained the scheme in his treatise the *Mysterium Cosmographicum*, published in 1596, in which his values for the relative radii of the planetary orbits agree reasonably well with those determined by Copernicus, though there was no physical basis for his theory.

The earth's orbit is the measure of all things; circumscribe around it a dodecahedron, and the circle containing it will be

Mars; circumscribe around Mars a tetrahedron, and the circle containing this will be Jupiter; circumscribe around Jupiter a cube, and the circle containing this will be Saturn. Now inscribe within the earth an icosahedron, and the circle contained in it will be Venus; inscribe within Venus an octahedron, and the circle contained in it will be Mercury. You now have the reason for the number of planets.

Kepler sent copies of his treatise to a number of scientists, including Galileo Galilei (1564–1642). In his letter of acknowledgment, dated 4 August 1597, Galileo congratulated Kepler for having had the courage, which he himself lacked, to publish a work supporting the Copernican theory.

Kepler wrote back to Galileo on 13 October 1597, encouraging him to continue supporting the Copernican theory. "Have faith, Galilii, and come forward!" he wrote. "If my guess is right, there are but few of the prominent mathematicians of Europe who would wish to secede from us: such is the power of truth."

Galileo was born in Pisa on 15 February 1564 to a Florentine family; they moved back to Florence in 1574. He enrolled in the school of medicine at the University of Pisa in 1581, studying physics and astronomy under Francesco Buonamici, who based his teachings on Aristotle. Galileo left Pisa without a degree in 1585 and returned to Florence, where he began an independent study of Euclid and Archimedes under Ostilio Ricci.

In 1583 Galileo made his first scientific discovery, that the period of a pendulum is independent of the angle through which it swings, at least for small angles. Three years later he invented a hydraulic balance, which he described in his first scientific publication, *La Balancetta* (The Little Balance), based on Archimedes' principle, which he also used in determining the centers of gravity of solid bodies.

Galileo was appointed professor of mathematics in 1589 at the University of Pisa, where he remained for only three years. During this period he wrote an untitled treatise on motion now referred to as *De Motu* (On Motion), which remained unpublished during his lifetime. The treatise was an attack on Aristotelian physics, such as the notion that heavy bodies fall more rapidly than light ones, which Galileo is supposed to have refuted by dropping weights from the leaning tower of Pisa. Through his study of balls rolling down an inclined plane, he found that the distance traveled was proportional to the square of the elapsed time, one of the basic laws of kinematics. He also concluded that a ball rolling on a frictionless horizontal surface would continue to roll with constant velocity, while one at rest would remain motionless, thus stating the law of inertia.

In 1592 Galileo was appointed to the chair of mathematics at the University of Padua, where he remained for eighteen years. During that period he wrote several treatises for the use of his students, including one that was first published in a French translation in 1634 under the title *Le Meccaniche*, a study of motion and equilibrium on inclined planes that further developed the ideas he had presented in *De Motu*.

In May 1597 Galileo wrote to a former colleague at Pisa defending the Copernican theory. Three months later he received a copy of *Mysterium Cosmographicum*, which led to his first correspondence with Kepler.

Kepler had also sent a copy of the *Mysterium Cosmographicum* to Tycho Brahe, who received it after he had left Denmark for Germany. Tycho responded warmly, calling the treatise "a brilliant speculation," beginning a correspondence that eventually led Kepler to accept Tycho's invitation to join him at his new observatory outside of Prague. As Tycho wrote in response to Kepler's letter of acceptance: "You will come not so much as a guest but as a very welcome friend and highly desirable participant and companion in our observations of the heavens."

Kepler finally arrived in Prague with his family early in 1600, beginning a brief but extraordinarily fruitful collaboration with Tycho. When Kepler began work at Prague he had hopes that he could take Tycho's data and use it directly to check his own planetary theory. But he was disappointed to find that most of Tycho's data was still in the form of raw observations, which first had to be subjected to mathematical analysis. Moreover, Tycho was extremely possessive of his data and would not reveal any more of it than Kepler needed for his work.

These and other disagreements with Tycho led Kepler to leave Prague in April of that year, though he returned in October after considerable negotiation concerning the terms of his employment. Tycho then assigned Kepler the task of analyzing the orbit of Mars, which until that time had been the responsibility of his assistant Longomontanus, who had just resigned. Kepler later wrote, "I consider it a divine decree that I came at exactly the time when Longomontanus was busy with Mars. Because assuredly either through it we arrive at the knowledge of the secrets of astronomy or else they remain forever concealed from us." Mars and Mercury are the only visible planets with eccentricities large enough to make their orbits significantly different from perfect circles. But Mercury is so close to the sun that it is difficult to observe, leaving Mars as the ideal planet for checking a mathematical theory, which is why Kepler was so enthusiastic about being able to analyze its orbit.

Early in the autumn of 1601 Tycho brought Kepler to the imperial court and introduced him to the emperor Rudolph II. Tycho then proposed to the emperor that he and Kepler compile a new set of astronomical tables. With the emperor's permission, this would be named the *Rudolfine Tables,* and since it was to be based on Tycho's observations it would be more accurate than any done in the past. The emperor graciously consented and agreed to pay Kepler's salary in this endeavor.

Soon afterward Tycho fell ill, and after suffering in agony for eleven days, on 24 October 1601, he died. On his deathbed he made Kepler promise that the *Rudolfine Tables* would be completed, and he expressed his hopes that it would be based on the Tychonic planetary model. As Kepler later wrote of Tycho's final conversation with him: "Although he knew I was of the Copernican persuasion, he asked me to present all my demonstrations in conformity with his hypothesis."

Two days after Tycho's death Emperor Rudolph appointed Kepler as court mathematician and head of the observatory in Prague. Kepler thereupon resumed his work on Mars, now with unrestricted access to all of Tycho's data. At first he tried the traditional Ptolemaic methods—epicycle, eccentric, and equant—but no matter how he varied the parameters the calculated positions of the planet disagreed with Tycho's observations by up to 8 minutes of arc. His faith in the accuracy of Tycho's data led him to conclude that the Ptolemaic theory of epicycles, which had been used by Copernicus, would have to be replaced by a completely new theory, as he wrote: "Divine Providence granted us such a diligent observer in Tycho Brahe, that his observations convicted this Ptolemaic calculation of an error of eight minutes; it is only right that we should accept God's gift with a grateful mind. . . . Because those eight minutes could not be ignored, they alone have led to a total reformation of astronomy."

After eight years of intense effort, Kepler was finally led to what are now known as his first two laws of planetary motion. The first law is that the planets travel in elliptical orbits, with the sun at one of the two focal points of the ellipse. The second law states that a radius vector drawn from the sun to a planet sweeps out equal areas in equal times, so that when the planet is close to the sun it moves rapidly and when far away it goes slowly. These two laws, which appeared in Kepler's *Astronomia Nova* (The New Astronomy), published in 1609, became the basis for his subsequent work on the *Rudolfine Tables*.

Kepler's first two laws of planetary motion eliminated the need for the epicycles, eccentrics, and equants that had been used by astronomers from Ptolemy to Copernicus. The passing of this ancient cosmological doctrine was noted by Milton in Book VIII of *Paradise Lost*, where he describes the debate between the two world systems, Ptolemaic and Copernican.

> *Hereafter, when they come to model Heaven,*
> *And calculate the stars; how they will wield*
> *The mighty frame; how build, unbuild, contrive*
> *To save appearances; how gird the sphere*
> *With centric and eccentric scribbled o'er,*
> *Cycle and epicycle, orb in orb.*

Kepler wrote two other works on his researches before the publication of his *Astronomia Nova*. The first was the *Appendix to Witelo*, published in 1604, which dealt with optical phenomena in astronomy, particularly

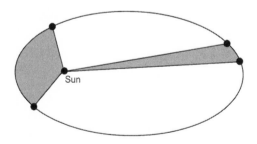

Kepler's first two laws of planetary motion. First law: the planets move in elliptical paths, with the sun at one of the two focal points of the ellipse. Second law: a radius vector drawn from the sun to a planet sweeps out equal areas in equal times.

parallax and refraction, as well as the annual variation in the size of the sun. The second book was occasioned by another new star, which appeared in October 1604 in the vicinity of Jupiter, Saturn, and Mars. Kepler published an eight-page tract on the new star in 1606 entitled *De Stella Nova*, with a subtitle describing it as "a book full of astronomical, physical, metaphysical, meteorological, astrological discussions, glorious and unusual." At the end of the tract Kepler speculated on the astrological significance of the new star, saying that it might be a portent of the conversion of the American Indians, the downfall of Islam, or even the second coming of Christ.

Meanwhile, the whole science of astronomy had been profoundly changed by the invention of the telescope. Instruments called "perspective glasses" had been used in England before 1580 for viewing distant terrestrial objects, and both John Dee and Thomas Digges were known to be expert in their construction and use, though there is no evidence that they used them for astronomical observations. But their friend Thomas Harriot, the Wizard Earl, is known to have made astronomical observations in the winter of 1609–10 with a small "telescope," which may have been a perspective glass.

Other than these perspective glasses, one of the earliest telescopes seems to have appeared in 1604, when a Dutch optician named Zacharias Janssen constructed one from a specimen belonging to an unknown Italian, after which he sold some of them at fairs in northern Europe. Hearing of the telescope, Galileo built one in his workshop in 1609 and then offered it to the Doge of Venice for use in war and navigation. After improving on his original design, he began using his telescope to observe the heavens, and in March 1610 he published his discoveries in a little book called *Siderius Nuncius* (The Starry Messenger).

The book begins with his observations of the moon, which he found to look very much like the earth, with mountains, valleys, and what he thought were seas. Seen in the telescope, the planets were pale illuminated disks, whereas the stars remained brilliant points of light. The Milky Way proved to consist of numerous stars, not a nebula reflecting the light of the sun, as some had thought, nor an atmospheric phenomenon, as Aristotle had concluded. He counted more than ninety stars in Orion's belt, where only nine are visible to the naked eye. He discovered four moons orbiting around Jupiter, a solar system in miniature, which he used as an additional argument in favor of the Copernican theory. He called the Jovian moons the "Medicean Stars" in honor of Cosimo de' Medici, the Grand Duke of Tuscany. Cosimo responded by making Galileo his court philosopher and appointing him to the chair of mathematics at the University of Pisa. Galileo had no obligation to teach at the University of Pisa or even to reside in the city, and so after his appointment, in September 1610, he departed to take up residence in Florence.

Galileo sent a copy of the *Siderius Nuncius* to Kepler, who received it on 8 April 1610. During the next eleven days Kepler composed his response in a little work called *Dissertatio cum Nuncio Sidereal* (Answer to the Starry Messenger), in which he expressed his enthusiastic approval of Galileo's discoveries and reminded readers of his own work on optical astronomy, as well as speculating on the possibility of inhabitants on the moon and arguing against an infinite universe.

Kepler borrowed a telescope from the elector Ernest of Cologne at the end of August 1610, and for the next ten days he used it to observe the heavens, particularly Jupiter and its moons. His excitement over the possibilities of the new instrument was such that he spent the next two months making an exhaustive study of the passage of light through lenses, which he published later in 1610 under the title *Dioptrice,* which became one of the foundation stones of the new science of optics.

The death of Rudolph II in early 1612 forced Kepler to leave Prague and take up the post of district mathematician at Linz, where he remained for the next fourteen years. One of his official duties was a study of chronology, part of a program of calendar reform instituted by Archduke Ferdinand II, son of the late emperor Rudolph. As a result of his studies he established that Christ was born in what in the modern calendar would be 5 B.C.

During that period when Kepler lived in Linz he continued his calculations on the *Rudolfine Tables* and published two other major works, the first of which was the *Harmonice Mundi* (Harmony of the World), which appeared in 1619. The title of this work was inspired by a Greek manuscript of Ptolemy's treatise on musical theory, the *Harmonica,* which Kepler acquired in 1607 and used in his analysis of music, geometry, astronomy, and astrology. The most important part of the *Harmonice Mundi* is the relationship now known as Kepler's third law of planetary motion, which he discovered on 15 May 1618 and presented in Book V. The law states that for each of the planets the square of the period of its orbital motion is proportional to the cube of its distance from the sun (or, strictly speaking, the semimajor axis of its elliptical orbit).

There had been speculations about the relation between the periods of planetary orbits and their radii since the times of Pythagoras, Plato, and Aristotle, and Kepler was terribly excited that he had at last, following in the footsteps of Ptolemy, found the mathematical law "necessary for the contemplation of celestial harmonies." He wrote of his pleasure, "That the same thought about the harmonic formulation had turned up in the minds of two men (though lying so far apart in time) who had devoted themselves entirely to contemplating nature ... I feel carried away and possessed by an unutterable rapture over the divine spectacle of the heavenly harmony."

Kepler dedicated the *Harmonice* to James I of England. The king responded by sending his ambassador Sir Henry Wooton with an invitation for Kepler to take up residence in England. But after considering the offer for a while Kepler eventually decided against it.

The English poet John Donne was familiar with the work of Copernicus and Kepler, probably through Thomas Harriot. Donne had in 1611 said to the Copernicans that "those opinions of yours may very well be true ... creeping into every man's mind." That same year Donne lamented the passing of the old cosmology in "An Anatomy of the World":

> And new Philosophy cals all in doubt,
> The Element of fire is quite put out;
> The Sun is lost, and th'earth, and no man's wit
> Can well direct him, where to look for it.

Kepler's second major work at Linz was his *Epitome Astronomiae Copernicanae* (Compendium of Copernican Astronomy), published in 1621. In the first three of the seven books of the *Epitome* Kepler refutes the traditional arguments against the motions of the earth, going much further than Copernicus and using principles that Galileo would later give in greater detail. His three laws of planetary motion are explained in great detail in Book IV, along with his lunar theory. The last three books treat practical problems involving his first two laws of planetary motion as well as his theories of lunar and solar motion and the precession of the equinoxes.

In 1626 Kepler was forced to leave Linz, which had undergone a two-month siege during a peasant uprising, and move to Ulm, where he published the *Rudolfine Tables* in September 1627, dedicating them to Archduke Ferdinand II. The new tables were far more accurate than any in the past, and they remained in use for more than a century. Kepler used his tables to predict that Mercury and Venus would make transits across the disk of the sun in 1631. The transit of Venus was not observed in Europe because it took place at night. The transit of Mercury was observed by Pierre Gassendi in Paris on 7 November 1631, representing a triumph for Kepler's astronomy, for his prediction was in error by only 10 minutes of arc as compared to 5 degrees for tables based on Ptolemy's model.

But Kepler did not live to see his theories vindicated, for he passed away on 15 November 1630. His tombstone, now lost, was engraved with an epitaph that he had written himself:

> *I used to measure the heavens,*
> *Now I measure the shadow of the earth.*
> *Although my soul was from heaven,*
> *The shadow of my body lies here.*

Meanwhile, Galileo had been active in advancing the cause of Copernicanism against the accepted cosmology of Aristotle, which in its reinterpretation by Saint Thomas Aquinas formed part of the philosophical basis for Roman Catholic theology. At the beginning of March 1616 the Holy Office of the Inquisition in Rome placed the works of Copernicus and all other writings that supported it, including those of Kepler, on the Index, the list of books that Catholics were forbidden to read. The decree held that believing the sun to be the immovable center of the world is "foolish and absurd, philosophically false and formally heretical." Pope Paul V instructed Cardinal Bellarmine to censure Galileo, admonishing him not to hold or defend Copernican doctrines any longer. On March 3 Bellarmine reported that Galileo had acquiesced to the pope's warning, and that ended the matter for the time being.

After his censure Galileo returned to his villa at Arcetri, outside Florence, where for the next seven years he remained silent. But in 1623, after the death of Gregory XV, Galileo became hopeful when he learned that his friend Maffeo Cardinal Barbarini had succeeded as Pope Urban VIII. Heartened by his friend's election, Galileo immediately proceeded to publish a treatise entitled *Il Saggiatore* (The Assayer), which appeared later that year, dedicated to Urban VIII.

Il Saggiatore grew out of a dispute over the nature of comets between Galileo and Father Horatio Grassi, a Jesuit astronomer. This had been stimulated by the appearance in 1618 of a succession of three comets, the third and brightest of which remained visible until January 1619. Grassi, who supported the Tychonic model of planetary motion, took the Aristotelian view that the comets were atmospheric phenomena, while Galileo insisted that they were in the celestial region. *Il Saggiatore* was favorably received in the Vatican, and Galileo went to Rome in the spring of 1623 and had six audiences with the pope. Urban praised the book, but he refused to rescind the 1616 edict against the Copernican theory, though he said that if it had been up to him the ban would not have been imposed. Galileo did receive Urban's permission to discuss Copernicanism in a book, but only if the Aristotelian-Ptolemaic model was given equal and impartial attention.

Encouraged by his conversations with Urban, Galileo spent the next six years writing a book called the *Dialogue Concerning the Two Chief World Systems, Ptolemaic and Copernican,* which was completed in 1630 and finally published in February 1632. The book is divided into four days of conversations among three friends: Salviati, a Copernican; Sagredo, an intelligent sceptic who had been converted to Copernicanism; and Simplicio, an Aristotelian.

The first day is devoted to a refutation of the Aristotelian view of the universe. On the second day the objections against the earth's motions are refuted on physical grounds. Many of Galileo's arguments here, though persuasive, are based on his erroneous notion of the inertia of circular motion. The third day is concerned with arguments for and against Copernicanism. Here, in comparing the two world systems, Galileo is often unfair in his criticism and exaggerates his claims for the superiority of the heliocentric theory. The fourth day is devoted to Galileo's erroneous theory of tidal action, which he believed to be conclusive proof of the earth's rotation.

Despite these defects, the arguments for Copernicanism were very persuasive and poor Simplicio, the Aristotelian, is defeated at every turn. Simplicio's closing remark represents Galileo's attempt to reserve judgment in the debate; he says, "It would still be excessive boldness for anyone to limit and restrict the Divine power and wisdom to some particular fancy of his own." This statement was apparently almost a direct quote of what Pope Urban had said to Galileo in 1623. When Urban read the *Dialogue* he remembered these words and was deeply offended, feeling that Galileo had made a fool of him and taken advantage of their friendship to violate the 1616 edict against teaching Copernicanism. The Florentine ambassador Francesco Niccolini reported that after he'd discussed the *Dialogue* with Urban, the pope broke out in great anger and fairly shouted, "Your Galileo has ventured to meddle with things that he ought not, and with the most grave and dangerous subjects that can be stirred up these days."

Urban directed the Holy Office to consider the affair and summoned Galileo to Rome. Galileo arrived in Rome in February 1633, but his trial before the court of the Inquisition did not begin until April. There he was accused of having ignored the 1616 edict of the Holy Office not to teach Copernicanism. The court deliberated until June before giving its verdict, and in the interim Galileo was confined in the palace of the Florentine ambassador. He was then brought once again to the Holy Office, where he was persuaded to acknowledge that he had gone too far in his support of the Copernican "heresy," which he now abjured. He was thereupon sentenced to indefinite imprisonment and his *Dialogue* placed on the Index. The sentence of imprisonment was immediately commuted to allow him to be confined in one of the Roman residences of the Medici family, after which he was moved to Siena and then, in April 1634, allowed to return to his villa at Arcetri.

Once Galileo returned home he took up again the researches he had abandoned a quarter of a century earlier, principally the study of motion. This gave rise to the last and greatest of his works, *Discourses and Mechanical Demonstrations Concerning Two New Sciences, of Mechanics and of Motions*, which he dictated to his disciple Vincenzo Viviani. The work was completed in 1636, when Galileo was seventy-two and suffering from failing eyesight. Since publication in Italy was out of the question because of the papal ban on Galileo's works, his manuscript was smuggled to Leyden, where the *Discourses* was published in 1638, by which time he was completely blind.

The *Discourses* is organized in the same manner as the *Dialogue*, divided into four days of discussions among three friends. The first day is devoted to subjects that Galileo had not resolved to his satisfaction, particularly his speculations on the atomic theory of matter. The second day was taken up with one of the two new sciences, now known in mechanical engineering studies as "strength of materials." The third and fourth days were devoted to the second of the two new sciences, kinematics, the mathematical description of motion, including motion at constant velocity; uniformly accelerated motion, as in free fall; nonuniformly accelerated motion, as in the oscillation of a pendulum; and two-dimensional motion, as in the parabolic path of a projectile.

Galileo died at Arcetri on 8 January 1642, thirty-eight days before what would have been his seventy-eighth birthday. The Grand Duke of Tuscany sought to erect a monument in his memory, but he was advised not to do so for fear of giving offense to the Holy Office, since the pope had said that Galileo "had altogether given rise to the greatest scandal throughout Christendom."

After Galileo's death a note in his hand was found on the preliminary leaves of his own copy of the *Dialogue*. He probably wrote it after the Holy Office imprisoned him for supporting the Copernican "heresy."

> Take note theologians, that in your desire to make matters of faith and of proposition relating to the fixity of sun and earth you may run the risk of eventually having to condemn as heretics those who would decide the earth to stand still and the sun to change position—eventually, I say—at such a time as it might be physically or logically proved that the earth moves and the sun stands still.

15

THE SCIENTIFIC REVOLUTION

The observations and theories of Copernicus, Tycho Brahe, Kepler, and Galileo, together with those of some of their contemporaries, represent the first stage of an intellectual upheaval that came to be called the Scientific Revolution, which continued through the seventeenth century and on into the early years of the eighteenth. Some historians today take issue with the concept of a "scientific revolution," a term coined in 1939, but all agree that it was a period during which the cosmology of western Europe changed profoundly and the modern scientific tradition emerged.

Two different philosophies of science were formulated in the seventeenth century. One was the empirical, inductive method of Francis Bacon (1561–1626); the other was the theoretical, deductive approach of René Descartes (1596–1650).

According to Bacon, the new science should be based primarily on observation and experiment, and it should arrive at general laws only after a careful and thorough study of nature. Bacon never accepted the Copernican theory, which he called a "hypothesis," and he criticized both Ptolemy and Copernicus for presenting nothing more than "calculations and predictions" rather than "philosophy . . . what is found in nature herself, and is actually and really true."

Descartes sought to give physical laws the same certitude as those of mathematics, a program that he laid out in his *Discourse on the Method of*

Reasoning Well and Seeking Truth in the Sciences. Whereas in philosophy Descartes began with the existence of self ("*Cogito ergo sum*," I am thinking, therefore I exist), in physics he started with the existence of matter, its extension in space, and its motion through space. That is, everything in nature can be reduced to matter in motion. Matter exists in discrete particles that collide with one another in their ceaseless motions, changing their individual velocities in the process, but with the total "quantity of motion" in the universe remaining constant. Descartes writes of the divine origin of this law in *The Principles of Philosophy* (1644). Speaking of God, he says, "In the beginning, in his omnipotence, he created matter, along with its motion and rest, and now, merely by his regular concurrence, he preserves the same amount of motion and rest in the material universe as he put there in the beginning."

Descartes presented his method in *Rules for the Direction of the Mind*, completed in 1628 but not published until after his death, and in the *Discourse on Method*, published in 1637 along with appendices entitled "Optics," "Geometry," and "Meteorology." He gave the final form of his three laws of nature in *The Principles of Philosophy*. The first law, the principle of inertia, states that "Each and every thing, insofar as it can, always continues in the same state, and thus what is once in motion always continues to move." The second law states that "all motion is in itself rectilinear . . . every piece of matter, considered in itself, always tends to continue moving, not in any oblique path but only in a straight line." The third law is concerned with collisions: "If a body collides with another body that is stronger than itself, it loses none of its motion; but if it collides with a weaker body, it loses a quantity of motion equal to that which it imparts to the other body."

The "Optics" presents Descartes's mechanistic theory of light, which he conceived of as a series of impulses propagated through the finely dispersed microparticles that fill the spaces between macroscopic bodies, leaving no intervening vacuum. This model gave him the right form for the law of refraction, but in his derivation he took the velocity of light to be greater in water than in air, which is not true.

The first correct derivation of the law of refraction appears to be the work of the Dutch mathematician Willebrord Snell (1580–1626), which was not published until after his death; it solved another problem that had eluded physicists since antiquity. The law, in its modern form, states that the ratio of the sines of the angles of incidence and refraction equals the ratio of the velocities of light in the two media.

The "Geometry" was inspired by what Descartes called the "true mathematics" of the ancient Greeks, particularly Pappus and Diophantus. Here he provided a geometric basis for algebraic operations, which to some extent had already been done by his predecessors as far back as al-Khwarizmi. The symbolic notation used by Descartes quickly produced great progress in algebra and other branches of mathematics. His work gave rise to the branch of mathematics now known as analytic geometry, which had been anticipated by Pierre Fermat (1601–65). Fermat, inspired by Diophantus and Apollonius, was also one of the founders of modern number theory and probability theory.

Descartes's "Meteorology" includes his theory of the rainbow, in which he used the laws of reflection and refraction to obtain the correct values of the angles at which the primary and secondary bows appear. He attempted to give a qualitative explanation of the colors of the rainbow, but this was based on his erroneous notion that light travels faster in water than in air.

Chapters 8 through 12 of Descartes's *Le monde, ou Traité de la lumière*, present his mechanistic cosmology, based on his theory of matter and laws of motion. This hypothetical "new world" that he describes consisted of an indefinite number of contiguous vortices, each with a star like our sun at the center of a planetary system, all carried around by the motion of the particles of the three types of matter that he believed filled all of space.

Descartes's vortex theory was generally accepted at first, but the researches of Christiaan Huygens (1629–95) showed conclusively that it was completely incorrect. Huygens was led to his rejection of the vortex theory by his studies of dynamics. In one of his studies he considered a situation in which a lead ball is attached to a string held by a man standing at the center of a rotating platform. When the platform rotates the man feels an outward or centrifugal force in the string attached to the ball, which in turn experiences an inward or centripetal force due to the string. Huygens found that the centripetal force on the ball was directly proportional to the mass of the ball and the square of its velocity and inversely proportional to the radius of its circular path, thus establishing the basis of dynamics for circular motion. This and his researches on the laws of collisions were what led Huygens to conclude that the Cartesian cosmology was in error. As he said in 1693, he could find "almost nothing I can approve as true in all the physics and metaphysics" of Descartes.

Descartes's Aristotelian notion that a vacuum was impossible was also shown to be incorrect by several of his contemporaries, beginning with Evangelista Torricelli (1608–47) and Blaise Pascal (1623–62). Torricelli's invention of the barometer in 1643 led him to conclude that the closed space above the mercury column represented at least a partial vacuum, and that the difference in the height of the two columns in the U-tube was a measure of the weight of a column of air extending to the top of the atmosphere. Pascal had a barometer taken to the top of the Puy de Dôme, a peak in central France, and it was observed that the difference in height of the two columns was less than at sea level, verifying Torricelli's conclusions. The results of this experiment led Pascal to urge all disciples of Aristotle to see if the writings of their master could explain the results. "Otherwise," he wrote, "let them recognize that experiments are the real masters that we should follow in physics; that the experiment done in the mountains overturns the universal belief that nature abhors a vacuum."

The German engineer Otto von Guericke (1602–1686) discovered that it was possible to pump air as if it were water, allowing him to produce a vacuum mechanically. In a famous experiment at Magdeburg in 1657, he pumped the air out of a spherical cavity made by fitting together two copper hemispheres, and showed that the resulting differential pressure was so great that not even two teams of horses pulling in opposite directions could force the two halves of the sphere apart.

Guericke's demonstration led the Irish chemist Robert Boyle (1627–1691) to make his own vacuum pump, using a design by the English physicist Robert Hooke (1635–1703). Boyle used his vacuum pump to do research on pneumatics, which he published in 1660 under the title *New Experiments Physico-Mechanicall, Touching the Spring of Air and Its Effects*. His conclusions were that a vacuum can be produced, or at least a partial one; that sound does not propagate in a vacuum; that air is necessary for life or a flame; and that air is elastic. In an appendix to the second edition of this work, published in 1662, he established the relationship now known as Boyle's law, that the pressure exerted by a gas is inversely proportional to its volume.

Boyle was influenced by both Francis Bacon's empiricism and Descartes's mechanistic view of nature. He was also influenced by the natural philosophy of Epicurus, which was revived by Pierre Gassendi (1592–1655), a French Catholic priest who in 1647 published a work in which he attempted to reconcile the atomic theory with Christian doctrine. This led Boyle to adopt a divinely ordained corpuscular version of mechanism, which he described in his treatise *Some Thoughts About the Excellence and Grounds of the Mechanical Philosophy*, published in 1670.

The culmination of the scientific revolution came with the career of Isaac Newton (1642–1727), whose supreme genius made him the central figure in the emergence of modern science.

Newton was born on 25 December 1642, the same year that Galileo died. His birthplace was the manor house of Woolsthorpe in Lincolnshire, England. His father, an illiterate farmer, had died three months before Isaac was born, and his mother remarried three years later, though she was widowed again after eight years. When Newton was twelve he was enrolled in the grammar school at the nearby village of Grantham, and he studied there until he was eighteen. His maternal uncle, a Cambridge graduate, sensed that his nephew was gifted and persuaded Isaac's mother to send the boy to Cambridge, where he was enrolled at Trinity College in June 1661.

At Cambridge Newton was introduced to both Aristotelian science and cosmology as well as the new physics, astronomy, and mathematics of Copernicus, Kepler, Galileo, Fermat, Descartes, Huygens, and Boyle. In 1663 he began studying under Isaac Barrow (1630–77), the newly appointed Lucasian professor of mathematics and natural philosophy. Barrow edited the works of Euclid, Archimedes, and Apollonius and published his own works on geometry and optics, with the assistance of Newton.

By Newton's own testimony he began his researches in mathematics and physics late in 1664, shortly before an outbreak of plague closed the university at Cambridge and forced him to return home. During the next two years, his *anni mirabiles*, he says, he discovered his laws of universal gravitation and motion as well as the concepts of centripetal force and acceleration.

This indicates that Newton had derived the law for centripetal force and acceleration by 1666, some seven years before Huygens, though he did not publish it at the time. He applied the law to compute the centripetal acceleration at the earth's surface caused by its diurnal rotation, finding that it was less than the acceleration due to gravity by a factor of 250, thus settling the old question of why objects are not flung off the planet by its rotation. He computed the centripetal force necessary to keep the moon in orbit, comparing it to the acceleration due to gravity at the earth's surface, and found that they were inversely proportional to the squares of their distances from the center of the earth. Then, using Kepler's third law of planetary motion together with the law of centripetal acceleration, he verified the inverse square law of gravitation for the solar system. At the same time he laid the foundations for the calculus and formulated his theory for the dispersion of white light into its component colors. "All this was in the two plague years 1665 and 1666," he wrote, "for in those years I was in the prime of my age for invention & minded Mathematicks & Philosophy more than at any time since."

When the plague subsided Newton returned to Cambridge, arriving in the spring of 1667. Two years later he succeeded Barrow as Lucasian professor of mathematics and natural philosophy, a position he was to hold for nearly thirty years.

During the first few years after he took up his professorship Newton devoted much of his time to research in optics and mathematics. He continued his experiments on light, examining its refraction in prisms and thin glass plates as well as working out the details of his theory of colors. He also carried on with his chemical experiments; like many of his contemporaries, he was still influenced by the old notions of alchemy.

Newton's silence allowed Robert Hooke (1635–1703) to claim that he was the first to discover the inverse-square law of gravitational force. In November 1662 Hooke had been appointed as the first curator of experiments at the newly founded Royal Society in London, a position he held until his death in 1703. Hooke made many important discoveries in mechanics, optics, astronomy, technology, chemistry, and geology. He is remembered today principally for the relation known as Hooke's law, which states that the force necessary to stretch a spring is proportional to the extension of the spring, a concept that can be applied to study any simple harmonic motion.

Meanwhile, Newton continued his researches on light, and he succeeded in making a reflecting telescope that was a significant improvement on any of the refractors then in use. News of his invention leaked out and he was urged to exhibit it at the Royal Society in London, which was just then beginning to hold its formal weekly meetings. The exhibit was so successful that Newton was proposed for membership in the Royal Society, and on 11 January 1672 he was elected as a fellow.

As part of his obligations as a fellow, Newton wrote a paper on his optical experiments, which he submitted on 28 February 1672, to be read at a meeting of the society. The paper, subsequently published in the *Philosophical Transactions of the Royal Society*, described his discovery that sunlight is composed of a continuous spectrum of colors, which can be dispersed by passing light through a refracting medium such as a glass prism. He found that the "rays which make blue are refracted more than the red," and he concluded that sunlight is a mixture of light rays, some of which are refracted more than others. Furthermore, once sunlight is dispersed into its component colors it cannot be further decomposed. This meant that the colors seen on refraction are inherent in the light itself and are not imparted to it by the refracting medium.

The procedures described in the paper were characteristic of Newton's favored approach to any scientific investigation. Later, in a controversy arising out of his first paper, Newton described his scientific method:

> For the best and safest method of philosophizing seems to be, first to enquire diligently into the properties of things, and to establish these properties by experiment, and then to proceed more slowly to hypotheses for the explanation of them. For hypotheses should be employed only in explaining the properties of things, but not assumed in determining them, unless so far as they may furnish experiments.

Ironically, the paper was widely criticized by Newton's contemporaries for just the contrary reason: that it did not confirm or deny any general philosophy of nature; the mechanists objected that it was impossible to explain his findings on the basis of any mechanical principles. Others insisted that Newton's experimental findings were false, since they themselves could not find the phenomena he had reported. Newton replied patiently to each of these criticisms in turn, but after a time he began to regret ever having presented his work in public. To make matters worse, Hooke began to claim that Newton's telescope was far inferior to one he himself had made.

For these and other reasons, early in 1673, Newton offered his resignation to the Royal Society. The secretary, Henry Oldenburg, refused to accept it and persuaded him to remain. But in 1676, after a public attack by Hooke, Newton broke off almost all association with the Royal Society. That same year Hooke became secretary of the society and wrote a conciliatory letter in which he expressed his admiration for Newton. Referring to Newton's theory of colors, Hooke said that he was "extremely well pleased to see those notions promoted and improved which I long since began, but had not time to compleat."

Newton replied in an equally conciliatory tone, referring to Descartes's work on optics. "What Descartes did was a good step. You have added much several ways, and especially in taking the colours of thin plates into philosophical consideration. If I have seen further than Descartes, it is by standing on the sholders [*sic*] of Giants."

Despite these friendly sentiments, the two were never completely reconciled, and Newton maintained his silence. Nevertheless, they continued to communicate with each other, a correspondence that was to lead again and again to controversy, the bitterest dispute arising from Hooke's claim that he had discovered the inverse-square law of gravitation before Newton.

By 1684 others beside Hooke and Newton were convinced that the gravitational force was responsible for holding the planets in their orbits, and that this force varied with the inverse square of their distance from the sun. Among them were the astronomer Edmund Halley (1656–1742), a good friend of Newton's and a fellow member of the Royal Society. Halley made a special trip to Cambridge in August 1684 to ask Newton "what he thought the Curve would be that would be described by the Planets supposing the force of attraction towards the Sun to be reciprocal to the square of their distance from it." Newton replied immediately that it would be an ellipse, but he could not find the calculation, which he had done seven or eight years before. And so he was forced to rework the problem, after which he sent the solution to Halley that November.

By then Newton's interest in the problem had revived, and he developed enough material to give a course of nine lectures in the fall term at

Cambridge, under the title of De Motu Corporum (The Motion of Bodies). When Halley read the manuscript of De Motu he realized its immense importance, and he obtained Newton's promise to send it to the Royal Society for publication. Newton began preparing the manuscript for publication in December 1684 and sent the first book of the work to the Royal Society on 28 April 1686.

On May 22 Halley wrote to Newton saying that the society had entrusted him with the responsibility for having the manuscript printed. But he added that Hooke, having read the manuscript, claimed that it was he who had discovered the inverse-square nature of the gravitational force and thought that Newton should acknowledge this in the preface. Newton was very much disturbed by this, and in his reply to Halley he went to great lengths to show that he had discovered the inverse-square law of gravitation and that Hooke had not contributed anything of consequence.

The first edition of Newton's work was published in midsummer 1687 at Halley's expense, since the Royal Society had found itself financially unable to fund it. Newton entitled his work *Philosophicae Naturalis Principia Mathematica* (The Mathematical Principles of Natural Philosophy); it is referred to more simply as the *Principia*. The *Principia* begins with an ode dedicated to Newton by Halley. This is followed by a preface in which Newton outlines the scope and philosophy of his work:

> Our present work sets forth mathematical principles of natural philosophy. For the basic problem of philosophy seems to be to discover the forces of nature from the phenomena of motions, and then to demonstrate the other phenomena from these forces.... Then the motions of the planets, the comets, the moon, and the sea are deduced from these forces by propositions that are also mathematical. If only we could derive the other phenomena of nature from mechanical principles by the same kind of reasoning!

The introduction begins with a series of eight definitions, of which the first five are fundamental to Newtonian dynamics. The first effectively defines "quantity of matter," or mass, as being proportional to the weight density times volume. The second defines "quantity of motion," subsequently to be called "momentum," as mass times velocity. In the third definition Newton says that the "inherent force of matter," or inertia, "is the power of resisting by which every body, so far as it is able, perseveres in its state either of rest or of moving uniformly straight forward." The fourth states, "Impressed force is the action exerted upon a body to change its state either of resting or of uniformly moving straight forward." The fifth through eighth define centripetal force as that by which bodies "are impelled, or in any way tend, toward some point as to a center." As an example Newton offers the gravitational force of the sun, which keeps the planets in orbit.

As regards the gravity of the earth, he gives the example of a lead ball, projected from the top of a mountain with a given velocity and in a direction parallel to the horizon. If the initial velocity is made larger and larger, he says, the ball will go farther and farther before it hits the ground, and may go into orbit around the earth or even escape into outer space.

The definitions are followed by a "*Scholium*," a lengthy comment in which Newton explains his notions of absolute and relative time, space,

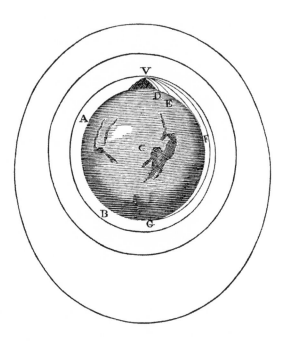

An illustration from Newton's *Principia* showing a projectile in orbit around the earth.

place, and motion. These essentially define the classical laws of relativity, which in the early twentieth century would be superceded by Einstein's theories of special and general relativity.

Next come the axioms, now known as Newton's laws of motion, three in number, each accompanied by an explanation and followed by corollaries:

> Law 1: Every body perseveres in its state of being at rest, or of moving uniformly forward, except insofar as it is compelled to change its state of motion by forces impressed. . . . Law 2: A change of motion is proportional to the motive force impressed and takes place along the straight line in which that force is impressed. . . . Law 3: To every action there is always an opposite and equal reaction; in other words, the action of two bodies upon each other are always equal, and always opposite in direction.

The first law is the principal of inertia, which is actually a special case of the second law when the net force is zero. The form used today for the second law is that the force F acting on a body is equal to the time rate of change of the momentum p, where p equals the mass m times the velocity v; if the mass is constant, then $F = ma$, a being the acceleration, the time rate of change of the velocity. The third law says that when two bodies interact the forces they exert on each other are equal in magnitude and opposite in direction.

Book 1 of the *Principia* is entitled "The Motion of Bodies." This begins with an analysis of motion in general, essentially using the calculus. First Newton analyzed the relations between orbits and central forces of various kinds. From this he was able to show that if and only if the force of attraction varies as the inverse square of the distance from the center of force, then the orbit is an ellipse, with the center of attraction at one focal point, thus proving Kepler's second law of motion. Elsewhere in Book 1 he proves Kepler's first and third laws. He also shows that, for inverse-square forces, the net force at a point within a spherical shell is zero, while the force outside a solid sphere is the same as if all the mass were concentrated at its center, so that in dealing with the solar system the sun and planets can be treated as point masses.

Book 2 is also entitled "The Motion of Bodies"; for the most part it deals with forces of resistance to motion in various type of fluids. One of Newton's purposes in this analysis was to see what effect the hypothetical *aether* in Descartes's cosmology would have on the motion of the planets. His studies showed that the Cartesian vortex theory was completely erroneous, for it ran counter to the laws of motion in resisting media that he'd established earlier in Book 2.

The third and final book of the *Principia* is entitled "The System of the World" and begins with three "Rules for the Study of Natural Philosophy." After this comes a section headed "Phenomena," in which he treats six of these, followed by forty-two "Propositions," each accompanied by a "Theorem" and sometimes followed by a "*Scholium*." This is in turn followed by a "General *Scholium*" and a concluding section entitled "The System of the World."

The six phenomena concern the motion of the planets and the earth's moon, along with observations about Kepler's second and third laws of planetary motion. He concludes that the planets "by radii drawn to the center . . . , describe areas proportional to the times, and their periodic times—the fixed stars being at rest—are as the $\frac{3}{2}$ powers of their distances from that center."

The first six propositions are arguments to show that the inverse-square gravitational force explains the motion of the planets orbiting the sun, the satellites of Jupiter, and the earth's moon, as well as the local gravity on the earth itself. The seventh proposition states Newton's law of universal gravitation: "Gravity exists in all bodies universally and is proportional to the quantity of matter in each."

Proposition 13 states Kepler's first and second laws of planetary motion: "The planets move in ellipses that have a focus in the center of the sun, and by radii drawn to that center they describe areas proportional to the times."

Proposition 18 says, "The axes of the planets are smaller than the diameters that are drawn perpendicular to the axes"—that is, the planets are oblate spheres. Newton correctly attributed this effect to the centrifugal forces arising from the axial rotation of the planets, so that the earth, for example, is flattened at the poles and bulges around the equator.

Proposition 24 presents Newton's theory of tidal action, that "the ebb and flow of the sea arises from the actions of the sun and moon," finally solving a problem that dated back to the time of Aristotle.

Proposition 39 is "To find the precession of the equinoxes," including the gravitational forces of both the sun and the moon on the earth. Newton correctly computed that "the precession of the equinoxes is more or less 50 seconds [of arc] annually," thus solving another problem that had preoccupied astronomers for some two thousand years.

Lemma 4 states, "The comets are higher than the moon, and move in the planetary regions." In the lemmas and propositions that follow, Newton discusses the motion of comets, showing that they move in elliptical orbits around the sun, thus reappearing periodically, as did the one known as Halley's comet, which had been observed in 1682 after disappearing seventy-five years before. He also speculated on the nature of comets, saying, as had Kepler, that the tail of a comet represents vaporization from the comet's head as it approaches the sun.

This is followed by a "General *Scholium*," in which Newton says that mechanism alone cannot explain the universe, whose harmonious order indicates to him the design of a Supreme Being. "This most elegant system of the sun, planets, and comets could not have arisen without the design and dominion of an intelligent and powerful being."

A second edition of the *Principia* was published in 1713 and a third in 1726, in both cases with a preface written by Newton. Meanwhile, in 1704 Newton had published his researches on light, much of which had been done early in his career. Unlike the *Principia*, which was in Latin, the first edition of his new work was in English, entitled *Opticks, or a Treatise of the Reflexions, Refractions, Inflexions and Colours of Light*. The first Latin edition appeared in 1706, and subsequent English editions appeared in 1717–18, 1721, and 1730; the last, which came out three years after Newton's death, bore a note stating that it was "corrected by the author's own hand, and left before his death, with his bookseller."

Like the *Principia*, the *Opticks* is divided into three books. At the very beginning of Book I Newton reveals the purpose he had in mind when composing his work. "My design in this Book," he writes, "is not to explain the Properties of Light by Hypotheses, but to propose and prove them by Reason and Experiment."

The topics dealt with in Book I include the laws of reflection and refraction, the formation of images, and the dispersion of light into its component colors by a glass prism. Other topics include the properties of lenses and Newton's reflecting telescope, the optics of human vision, the theory of the rainbow, and an exhaustive study of color. Newton's proof of the law of refraction is based on the erroneous notion that light travels more rapidly in glass than in air, the same error Descartes had made. This error stems from the fact that both of them thought that light was corpuscular in nature.

Newton's corpuscular view of light stemmed from his acceptance of the atomic theory. He writes of his admiration for "the oldest and most celebrated Philosophers of Greece . . . who made a Vacuum, and Atoms, and the Gravity of Atoms, the first Principles of their Philosophy." Later on, he says, "All these things being consider'd, it seems to me that God in the Beginning formed Matter in solid, hard, impenetrable, moveable Particles, of such Sizes and Figures, and with such other Properties and in such Proportions to Space, as much conduced to the End for which he had form'd them."

Book II begins with a section entitled "Observations Concerning the Reflexions, Refractions, and Colours of Thin Transparent Bodies." The effects he studied here are now known as interference phenomena; Newton's observations are the first evidence for the wavelike nature of light.

In Book II Newton also comments on the work of the Danish astronomer Olaus Roemer (1644–1710), who in 1676 measured the velocity of light by observing the time delays in successive eclipses of the Jovian moon Io as Jupiter receded from the earth. Roemer's value for the velocity of light was about a fourth lower than the currently accepted one of slightly less than 300,000 kilometers per second, but it was nevertheless the first measurement to give an order of magnitude estimation of one of the fundamental constants of nature. Roemer computed that light would take eleven minutes to travel from the sun to the earth, as compared to the correct value of eight minutes and twenty seconds. Newton seems to have made a better estimate of the speed of light than Roemer, for in Book II of the *Opticks* he says, "Light is propagated from luminous Bodies in time, and spends about seven or eight Minutes of an Hour in passing from the Sun to the Earth."

The opening section of Book III deals with Newton's experiments on diffraction. The remainder of the book consists of a number of hypotheses, not only on light, but on a wide variety of topics in physics and philosophy. The first edition of the *Opticks* had sixteen of these "Queries," the second twenty-three, the third and fourth thirty-one. It would seem that Newton, in the twilight of his career, was bringing out into the open some of his previously undisclosed speculations, his heritage for those who would follow him in the study of nature.

Meanwhile, Newton had been involved in a dispute with the great German mathematician and philosopher Gottfried Wilhelm Leibniz, the point of contention being which of them had been the first to develop the calculus. According to his own account, Newton first conceived the idea of his "method of fluxions" around 1665–66, although he did not publish it until 1687, when he used it in the *Principia*. He first published his work on the calculus independently in a treatise that came out in 1711. Leibniz began to develop the general methods of the calculus in 1675, though he did not publish his work until 1684. The version of calculus formulated by Leibniz, whose notation was much like that used today, caught on more rapidly than that of Newton, particularly on the Continent. Newton's bitterness over the dispute was such that in the third edition of the *Principia* he deleted all reference to Leibniz, who until the end of his days continued to accuse his adversary of plagiarism.

Aside from his work in science, Newton also devoted much of his time to studies in alchemy, prophecy, theology, mythology, chronology, and history. His most important nonscientific work is *Observations upon the Prophecies of Daniel, and the Apocalypse of St. John,* which is considered to be a possible key to the method of his alchemical studies, as evidenced by such notions as his analogy between the "four metals" of alchemy and the four beasts of the Apocalypse.

Newton died in London on 20 March 1727, four days after presiding over a meeting of the Royal Society, of which he had been president since 1703. His body lay in state until April 4, when he was buried with great pomp in Westminster Abbey. Voltaire, writing of Newton's funeral, noted, "He lived honored by his compatriots and was buried like a king who had done good to his subjects."

In the last days of his life Newton had remarked, "I do not know what I may appear to the world; but to myself I seem to have been only like a boy, playing on the seashore, and diverting myself, in now and then finding a smoother pebble or a prettier shell than ordinary, whilst the great ocean of truth lay all undiscovered before me."

SAMARKAND TO ISTANBUL:
THE LONG TWILIGHT OF
ISLAMIC SCIENCE

Newton paid tribute to his predecessors when he said that if he had seen farther than Descartes it was "by standing on the sholders of Giants." The colossal figures he was referring to can be identified from his works, where he gives credit to his European predecessors, most notably Copernicus, Tycho Brahe, Kepler, and Galileo, and to the ancient Greeks, including Pythagoras, Empedocles, Philolaus, Democritus, Plato, Aristotle, Epicurus, Euclid, Archimedes, Apollonius, Aristarchus, Diophantus, Ptolemy, and Pappus of Alexandria.

But Newton makes no mention of any Arabic scientists, though surely he must have been aware that much of Greek science had been transmitted to Europe through the Islamic world. Islamic science effectively had come to an end by the time the Scientific Revolution began, and the works of medieval Arabic philosophers, physicists, mathematicians, astronomers, engineers, astrologers, and alchemists had been for the most part either lost or forgotten. The science of ancient Greece and medieval Islam had been supplanted by the new world system that had emerged during the Scientific Revolution, and that in the two centuries after Newton would give rise to the Industrial Revolution and the Atomic Age.

Historians of science until the mid-twentieth century were of the opinion that Islamic science reached its peak in the late medieval period and then declined rapidly, just as European science was beginning to emerge. William Cecil Dampier, whose *History of Science and Its Relation with Philosophy and Religion* appeared in three editions and twelve reprints between 1929 and 1945, devotes only 7 of the 574 pages of his book to Islamic science. He writes of the "absorption of Arabic knowledge by the Latin nations" beginning at the close of the eleventh century, when, according the his view, "the decline of Arabic and Muslim learning had set in."

The decline of Arabic science was accelerated by the rise of the Mongols and their sack of Baghdad in 1258 under Hulagu, who executed the last Abbasid caliph. This was a turning point in the history of eastern Islam, for the Mongol invasion opened up the way for the westward migration of Turkish-speaking people from the steppes of central Asia. The Seljuk Turks were followed by the Ottoman Turks, who after their conquest of Constantinople in 1453 created an empire that extended from southern Europe through the Middle East and North Africa. The Turkish historian Aydín Adívar, writing in 1939, put forward the view that the Ottoman sultanate cut itself off from Western science, which did not reach Turkey and the Middle East until after the collapse of the empire and the creation of the modern Republic of Turkey in 1923. But recent research has shown that Islamic science reached a new peak under the Mongols two centuries after their sack of Baghdad, and that it continued at a high level for at least another century under the Ottoman Turks before beginning its inexorable decline.

After the fall of the Abbasid dynasty several important astronomical observatories were founded in Central Asia by Mongol-Turkic rulers. The three most renowned were at Maragha and Tabriz, in Persia, and at Samarkand, in what is now Uzbekistan. A number of Arabic astronomers, physicists, and mathematicians made important advances at these observatories during the two centuries following the Mongol conquest of Baghdad, which has led at least one modern historian to describe that era as the golden age of Islamic science.

The Maragha observatory was founded in 1259 by the Ilkhanid Mongol ruler Hulagu Khan, grandson of Genghis Khan. The first director of the observatory and its research center, which included a school of astronomy and a library, was the Persian astronomer and mathematician Nasir al-Din al-Tusi (1201–74). The actual construction of the observatory and its instruments was directed by the astronomer Mu'ayyad al-Din al-'Urdi (d. 1266) of Damascus.

Hulagu's deed of foundation gave the observatory financial independence, so that it survived his death in 1265 and continued in operation until 1316. During that time at least eighteen astronomers are known to have worked at Maragha, including one from North Africa and another from China. The instruments they used included a mural quadrant with a radius of more than sixty feet, graduated to read minutes of arc. These instruments were used by al-Tusi and his staff to compile the *Zij-i Ilkhani*, the Ilkhanid astronomical tables, which were completed in 1272 under Hulagu's successor, Abaqa Khan.

Al-Tusi also wrote a book for the general reader called *The Treasury of Astronomy*, which criticized Ptolemaic concepts such as the epicycle theory and introduced new planetary models. One of his innovations was the so-called al-Tusi couple, which has one circle rolling inside another to give a combination of two circular motions, a substitute for the Ptolemaic epicycles. The al-Tusi couple was used effectively by a number of his successors, both Arabic and European, up to and including Copernicus.

Besides his astronomical writings, al-Tusi, who was fluent in both Persian and Arabic, wrote numerous works on geometry, trigonometry, mineralogy, alchemy, astrology, philosophy, logic, ethics, and theology. His works on astronomy and mathematics were translated into Latin and were influential in the development of European science.

Mu'ayyad al-Din al-'Urdi wrote a treatise devoted to the reform of Ptolemaic astronomy. Entitled simply *A Book on Astronomy*, it apparently predates al-Tusi's work on the same subject, and is thus the first Arabic work to offer an alternative to the Ptolemaic epicycle theory. One of his mathematical methods, known as Urdi's lemma, was also used by subsequent astronomers up to and including Copernicus.

Two other notable astronomers at the Maragha observatory were Muhyi al-Din al-Maghribi (d. ca. 1290) and Qutb al-Din al-Shirazi (1236–1311).

As his last name indicates, Muhyi al-Din was born in the Maghreb, Muslim northwest Africa. He first studied religious law in the Maghreb and then moved to Aleppo, where he served as court astrologer to the Ayyubid sultan al-Nasir II. By his own testimony, he escaped death when the Mongols conquered Syria by simply telling them that he was an astrologer. He then went to work with Nasir al-Din al-Tusi at the Maragha observatory, where he remained for the rest of his days. His extant manuscripts include the *Compendium of the* Almagest and numerous other works in astronomy and mathematics, one of the latter being a commentary on the *Conics* of Apollonius.

Qutb al-Din al-Shirazi took his name from the Persian city of Shiraz, where his father, Mas'ud al-Qadharuni, was a renowned physician at the Muzaffari hospital. When Mas'ud died Qutb al-Din was only fourteen, but he was mature enough to take over his father's duties at the hospital, where he worked for the next ten years. He then went to Maragha to study astronomy and mathematics with Nasir al-Din al-Tusi, who later saw him as a rival and expelled him from the observatory. Al-Shirazi then went to Tabriz, where, under the patronage of the Ilkhanid Mongol ruler Ghazan Khan and his successor, Oljaytu, he founded an observatory that became the successor to the one at Maragha.

Al-Shirazi is credited with further developing the theory of the al-Tusi couple, which he had learned from al-Tusi at Maragha. His major astronomical work is *The Limit of Understanding of the Knowledge of the Heavens*, which, in addition to astronomy, has sections on mechanics, optics, meteorology, geography, geodesy, and cosmography. Another of his works is entitled *A Book I Have Composed on Astronomy, but Do Not Blame Me*. Al-Shirazi is also renowned for his medical writings, particularly his commentary on the *Canon* of Ibn Sina, which he defended against the attacks of the theologians.

Al-Shirazi's most brilliant student at Tabriz was Kamal al-Din al-Farisi (1267–1319), whose last name is derived from his birthplace, the Persian city of Fars. At al-Shirazi's suggestion, al-Farisi wrote a series of commentaries on the optical works of Ibn al-Haytham, which he then followed with his own treatise on the science of light, entitled *Revision of Optics*.

Al-Farisi made several advances on the researches of Ibn al-Haytham, most notably in his theory of the rainbow. Here he used a hollow glass sphere filled with water as an analogue for a raindrop. His studies led him to conclude that the rainbow is produced by a combination of refraction and internal reflection of sunlight in the individual drops of water suspended in the air after a rainfall. In the primary rainbow, according to his theory, the light enters the drop and is internally reflected once before leaving, undergoing refraction on entry and departure, while in the secondary bow there are two internal reflections. The colors are due to the refractions, with their order from red to blue inverted in the secondary bow due to the second internal reflection.

The Turkish historian Mustafa Nazif has concluded that al-Farisi published his theory of the rainbow at least a decade before Dietrich of Freiburg, whose researches on the same subject date to the years 1304–11 and led him to the same conclusion. Dietrich refers to the optical works of Ibn al-Haytham but he does not mention the writings of al-Farisi, which were never translated into Latin.

The observatory at Samarkand was founded in 1425 by the Timurid khan Ulugh Beg, grandson of the great Mongol ruler Tamerlane. The observatory was erected on the same site where Ulugh Beg had four years earlier built a madrasa, a college of theology to which he had added a school for advanced study in science and mathematics. He directed his foundation until 1449, when he was assassinated by his brother. Ulugh Beg's observatory closed a few years afterward, with a singular record of accomplishments despite its relatively brief lifetime.

The principal astronomer at the Samarkand observatory during its early years was Jamshid al-Kashi (d. 1429), from Kashan in northern Persia. Al-Kashi's principal astronomical work is the *Zij-i Khaqani*, a revision of the *Zij-i Ilkhani* of Nasr al-Din al-Tusi, to which he added trigonometric tables and descriptions of a number of different calendars that had been used in central Asia, including those of the Uigur Turks and the Ilkhanid Mongols. Another of his astronomical works, *The Stairway of Heaven*, is an attempt to measure the distance and sizes of the planets. Other treatises describe the astronomical instruments he used in his observations, some of them his own inventions.

Al-Kashi's best-known mathematical work is *Miftah al-Hisab*, an encyclopedia of elementary mathematics, used for centuries by astronomers as well as architects, surveyors, and merchants. He also wrote two other mathematical treatises in connection with his researches in astronomy, where his method of approximation in calculating precise trigonometric tables anticipates the work of later European mathematicians.

When al-Kashi died, in 1429, he was succeeded as chief astronomer by Qadi Zada al-Rumi (ca. 1364–ca. 1436). Qadi Zada was born and educated in Bursa, the first capital of the Ottoman Turks, in northwestern Asia Minor. He traveled to Samarkand and presented himself to Ulugh Beg, who in 1421 appointed him rector of his newly founded madrasa. After he became head of the observatory he wrote a number of treatises on astronomy and mathematics, including a commentary on the *Almagest* and a revision of Euclid's *Elements*.

When Qadi Zada died (ca. 1436), he was succeeded as head astronomer by Ali al-Qushji (ca. 1402–1474). Ali was born in Samarkand, and took the name of Qushji, or Bird-Man, because in his youth he was Ulugh Beg's falconer. He subsequently served as Ulugh Beg's ambassador to China. After becoming chief astronomer he supervised the completion of Ulugh Beg's astronomical tables, the *Zij-i Sultanai*, published in 1438. These tables were probably first written in Persian and soon afterward translated into Arabic and Turkish.

Al-Qushji left Samarkand soon after Ulugh Beg's death. He later went to Istanbul as chief astronomer for the Ottoman sultan Mehmet II (r. 1451–81), known as Fatih, or the Conqueror, in honor of his capture of Constantinople in 1453. Al-Qushji's writings include two treatises dealing with the solution of Ptolemaic models, one for the moon and the other for Mercury, as well as an introductory work called *The Fathiya Treatise on Astronomy*, dedicated to Mehmet II.

The first observatory in Istanbul was founded during the reign of Sultan Murat III (r. 1574–95) by Takiyuddin al-Rashid (1526–1585), an astronomer from Damascus. Takiyuddin equipped the observatory with several new instruments of his own invention, as well as a mechanical clock that he had made himself. His first project at the observatory was to correct Ulugh Beg's astronomical tables. At least one of his measurements was more accurate than that of Tycho Brahe. This was the annual motion of the sun's apogee in the celestial sphere, which he measured as 63 seconds of arc and which Tycho recorded as 45, compared to the currently accepted value of 61.

Takiyuddin also made careful observations of the comet of 1578, though he does not appear to have drawn the same conclusion as Tycho Brahe, who said that the fiery body was passing through the planetary celestial spheres. The poet Aladdin al-Mansur, in his poem "Concerning the Appearance of a Fiery Stellar Body," writes that the comet appeared

Takiyuddin's Istanbul Observatory.

on the first night of the holy month of Ramadan, "passing through the nine sections of the ephemeral world . . . / like a turban sash over the Ursa Minor stars."

Takiyuddin, who was also the court astrologer, saw the comet as a sign of good fortune and predicted that the Ottomans would gain victory in their war against the Persians. But the head of the Muslim religious hierarchy, the Sheikh ül-Islam Kadizade, convinced Sultan Murat that the observatory would bring disaster to the realm by prying into the secrets of nature, as evidenced by the fate of Ulugh Beg, who had been assassinated after his planetary tables were published, he argued. Aladdin al-Mansur writes of how the sultan questioned Takiyuddin: "People of learning have made inquiries concerning this. Oh you learned man of consciousness and perfection, inform me once more of the progress and the results of your observations. Have you entangled knots from the firmament in a hair-splitting manner?" Takiyuddin answered, "In the Zij of Ulugh Beg there were many doubtful points, oh exalted king. Now through observations the tables have been corrected, and out of grief the heart of the foe has withered and twisted into coils. From now on, order the abolishment of the Observatory."

Aladdin al-Mansur, in the last lines of his poem, describes the fate of the Istanbul observatory. "The King of Kings summoned the Head of the Halberdiers of his Bodyguard and gave him instructions for the demolishment and abolition of the Observatory. Orders were given that the Admiral should immediately rush to the Marine Ordnance Division and they should at once wreck the Observatory and pull it down from its apogee to its perigee."

During the reign of Sultan Selim III (r. 1789–1807), efforts to modernize the Ottoman army led to the establishment, in 1793, of a school for artillery officers, originally called the Mühendishane-i Cedide, or Military Engineering School. The curriculum included classes on mathematics, geography, and astronomy, as evidenced by the lecture notes of Hüseyin Rifki Tamani, head teacher during the years 1806–17. But Tamani still based his astronomy teaching on the old Ptolemaic model, as he remarks at the conclusion of one lecture: "Let it be known that the universe in appearance is a sphere and its center is the Earth. . . . The Sun and Moon rotate around the globe and move about the signs of the zodiac."

Ishak Efendi (1774–1836), who became head of the Mühendishane in 1830, wrote a four-volume survey of contemporary scientific knowledge in Europe, including the works of Descartes and Newton. The fourth volume included 257 pages on astronomy, in which he says that the Copernican theory can explain many astronomical events more easily than the old geocentric model of Ptolemy. The fourth volume of Ishak Efendi's work was first printed in 1834 in Istanbul; eleven years later it was reprinted in Cairo. During the last Ottoman century it was the principal source of knowledge in the empire for those interested in the new science that had been developed in western Europe.

The first attempt to establish an Ottoman institution of higher learning, the Darülfünun, in Turkish, was begun during the reign of Sultan Abdul Mecit (r. 1839–61), as part of the reform movement known as the Tanzimat. The Darülfünun, which registered its first students in 1869, was reorganized in 1900 on the model of American and European universities, including faculties of science and medicine. After the founding of the Republic of Turkey in 1923 the Darülfünun became the University of Istanbul and the old Mühendishane was reorganized as Istanbul Technical University.

Another scientific institution founded in the second half of the nineteenth century was the Rasathane-i Amiri, or Imperial Observatory, though its primary function was as a meteorological station. Turkish astronomers began making observations at the Rasathane in 1910, and soon after the establishment of the Republic of Turkey it was moved to its present site, at Kandilli on the Asian shore of the Bosphorus, also serving as a seismological research center as well as a weather station.

Today the Rasathane is attached to Bosphorus University (BU), established in 1971 in the buildings and grounds of the old Robert College (RC), which was founded as an American missionary school in 1863 on the European shore of the Bosphorus. I taught physics at RC-BU in the years 1960–76, and since 1993 I have been lecturing on astronomy and the history of science at BU, an institution that spans the gap between the Ottoman Empire and the Republic of Turkey, the old Turkey and the new.

One of the most important references in my history of science course these days is an encyclopedic work published in Istanbul in 2003 by Boris A. Rozenfeld and Ekmeleddin Ihsanoğlu titled *Mathematicians, Astronomers and Other Scholars of Islamic Civilisations and Their Works (7th–14th Centuries)*. Commonly known as *MASI*, the work is a survey of 1,711 scientists, whose manuscripts, along with 1,376 works whose authors are unknown, are preserved in the libraries of fifty countries. Most of the manuscripts are written in Arabic, but some are in Persian, Syriac, Sanskrit, Tajiki, Urdu, Old Turkish, Tatar, Uzbek, or other Asian languages. The subject headings under which the works are classified are mathematics, astronomy, mechanics, physics, music, mathematical geography, descriptive geography, chemistry and alchemy, mineralogy, meteorology, zoology, botany, philosophy and theology, literature and linguistics, and mysticism. The manuscript libraries listed in *MASI* include 29 in Iraq, 27 in Iran, 25 in Turkey, 15 in India, 10 in Egypt, 9 in Afghanistan, 8 each in Morocco and Russia, 6 each in Lebanon, Spain, and Syria, 5 each in Pakistan, Uzbekistan, and Yemen, 4 each in Tajikistan and Ukraine, 2 each in Algeria, Azerbaijan, Bosnia and Herzegovina, Portugal, Saudi Arabia, and Tunisia, and 1 each in Armenia, Bangladesh, Bulgaria, Georgia, Indonesia, Kazakhstan, Libya, Nigeria, Qatar, and Turkmenistan, to name only the countries that once had centers of Islamic science.

Sixteen of the manuscript libraries of Turkey listed in *MASI* are in Istanbul, including one at the Kandilli Rasathane, which is solely devoted to Ottoman astronomers. A number of my students in the history of science have written term papers on research they have done in the library at the Kandilli Rasathane, particularly on the observations of Takiyuddin, founder of the first observatory in Istanbul.

The richest collection of Islamic manuscripts in Istanbul is in the library of the Süleymaniye mosque complex, built in 1550–55 by Sultan Süleyman the Magnificent (r. 1520–66). The oldest of the manuscripts in the Süleymaniye library date back to the beginning of the Islamic renaissance in Baghdad. But since all of the manuscripts are either in Arabic, Old Turkish, or other languages indecipherable to me and my students, we are forced to rely on translations into modern Turkish or English, which has been done for only a very few works, or on scholarly surveys of Islamic science such as *MASI*.

The question my students always ask and try to answer is why Islamic astronomy declined so sharply after the time of Takiyuddin, whose contemporary Tycho Brahe paved the foundation for the new astronomy of western Europe. While trying to find an answer they are generally surprised to find that Islamic astronomers continued to scan the heavens long after the time of Takiyuddin. I tell them that it is because astronomy was one branch of science that was always accepted in Islam, since astronomers used their skills to determine the months of the Muslim calendar and the five times of daily prayer; to orient mosques toward the *qibla*, the direction of Mecca; to predict eclipses of the sun and moon; and to follow the motions of the planets so as to

236 | *Aladdin's Lamp*

draw up horoscopes for the caliphs, khans, emirs, and sultans who employed them. Aladdin al-Mansur writes of how Takiyuddin saved his skin by interpreting the comet of 1578 as a most favorable omen for Sultan Murat III, although he was forced to sacrifice his observatory in the process.

And so Islamic astronomers continued to make observations in the same way as had their Arabic and Greek predecessors, while their European contemporaries began the intellectual revolution that led to the emergence of modern science. Thus the Islamic world was left behind as its vast empires declined and fell, living on the fading memory of the great accomplishments of its physicists, physicians, mathematicians, geographers, and astronomers, who passed on Greek science to the West along with the advances they had made on their own.

17

SCIENCE LOST AND FOUND

Many of the great works of ancient science were lost in the collapse of Greco-Roman civilization, though in the past century a few of these classics have been rediscovered from the dustbin of history, in some cases almost miraculously.

In 1900 a Greek sponge boat from the island of Symi anchored off the northern coast of the remote islet of Antikythera to escape a storm. After the storm abated a diver named Elias Stadiatos went down to look for sponges and found a wrecked ship on the sea bottom. He was stunned by what he found, and talked excitedly of having seen a heap of dead naked women, which when they were brought to the surface were found to be Greco-Roman bronze statues. The other contents of the ship included jewelry, bronzes, pottery, furniture, and amphorae filled with wine. The ship was dated to the first century B.C. and is thought to have been on its way from Rhodes to Italy. One of the bronze statues, known as the Ephebe of Antikythera, is now in the National Archaeological Museum in Athens. It represents a nude youth thought to be Paris, the son of King Priam of Troy, possibly done by the renowned sculptor Euphranor, who worked in Athens in the mid-fourth century B.C.

Almost overlooked in the objects brought up from the wreck was a wooden box about the size of a book, which when opened proved to contain a complex arrangement of bronze gears and dials, all heavily eroded into shapeless lumps of green metal. The wooden box soon disintegrated into dust, but the bronze gears and dials survived and were eventually subjected to an X-ray analysis to determine their function. The device, now known as the Antikythera computer, proved to be an elaborate clockwork mechanism, which Derek De Solla Price, a historian of science at Yale, showed to be an astronomical device that reproduced the motions of the sun, moon, and visible planets. Such a device, now called an orrery, or planetarium, was known to the Greeks as a *sphairopoiia*, a mechanism that modeled the motions of the celestial bodies. A more recent analysis by Michael Wright, curator at the Science Museum in London, has shown that the gear system reproduced the epicycle theory for planetary motions derived by Apollonius of Perge and used by Ptolemy of Alexandria.

Several classical sources credit Archimedes with the invention of two mechanical devices in bronze that reproduced the motions of the heavenly bodies. One of these sources, the mathematician Pappus of Alexandria, says that Archimedes wrote a treatise, now lost, called *Peri Spairopoiias* (On Sphere-Making), describing a celestial globe that he made to represent the motions of the sun and moon and demonstrate both solar and lunar eclipses. The other Archimedean invention was an orrery that reproduced the motions of the planets as well as those of the sun and moon.

Cicero reports that the Roman general Marcellus took both of these Archimedean devices back to Rome as booty after his sack of Syracuse in 212 B.C. He set up the celestial globe in the temple of Vesta for all to see. The poet Ovid, writing circa A.D. 8, describes the globe in his verses on the goddess and her sanctuary: "There stands a globe hung by Syracusan art in closed air, a small image of the vast vault of heaven, and the Earth is equally distant from the top and bottom. That is brought about by its round shape."

The orrery eventually came into the possession of a grandson of Marcellus, who showed it to the astronomer Gaius Sulpicius Gallus. Gallus used the orrery to predict a lunar eclipse on 21 June 168 B.C., and Cicero says that he demonstrated solar eclipses as well, though obviously he could not predict whether these would be visible in Rome.

Other astronomers followed the lead of Archimedes in constructing these mechanical devices for demonstrating celestial motions, and examples survive from the Byzantine, Islamic, and late medieval European eras. Cicero writes, "Our friend Posidonius as you know has recently made a globe which in its revolution shows the movement of the sun and stars and planets, by day and night, just as they appear in the sky." Cicero would have seen this orrery, for as a young man he had attended the lectures of Posidonius at Rhodes. This suggests the possibility that the Antikythera computer may have been the orrery made by Posidonius, since the ship in which it was found is thought to have been heading from Rhodes to Italy.

The finding of the Antikythera computer was the first of two dramatic discoveries of lost works of Greek science made at the beginning of the twentieth century. The second came in 1906, when the Danish scholar John Ludwig Heiberg discovered a copy of a lost work of Archimedes' in a Greek church in Istanbul, along with other ancient manuscripts. Heiberg's remarkable finding was reported on the front page of *The New York Times* on 16 July 1907, though it failed to receive any mention in *The Times* of London.

Heiberg made his discovery in the church of Agios Giorgios (St. George) Metochi in the old Greek quarter of the Fener on the Golden Horn. Agios Giorgios is a metochion, or daughter church, of the Monastery of the Holy Sepulchre in Jerusalem and belongs to the Jerusalem Patriarchate rather than to the Patriarchate of Constantinople, which has had its headquarters in the Fener since the end of the sixteenth century. Among the manuscripts discovered there was one now known as Codex C, Archimedes' thesis *On the Method*, which had been lost for some two thousand years.

Newton and all of his European predecessors were aware of their great debt to Archimedes, who is mentioned more than a hundred times by Galileo, for his rigorously mathematical approach to the study of nature became the model for the new science that replaced the moribund Aristotelianism of the medieval era. As Marshall Clagett writes regarding the influence of Archimedes on Galileo and his contemporaries: "Archimedes' significance for these founders of early modern science lay in the use of mathematics in the treatment of physical problems as well as in the originality and fertility of his mathematical techniques."

Given the crucial importance of Archimedes in the Scientific Revolution, it is somewhat of a miracle that his extant writings survived at all.

Carl B. Boyer writes in his history of mathematics, "Unlike the *Elements* of Euclid, which have survived in many Greek and Arabic manuscripts, the treatises of Archimedes have reached us through a slender thread. Almost all copies are from a single Greek original which was in existence in the sixteenth century and itself copied from an original of about the ninth or tenth century." One of the most important of Archimedes' works, his treatise *On the Method*, was believed to have been lost in late antiquity; though its rediscovery by Heiberg caused a sensation, it disappeared from sight again a few years later. Finally, it reemerged at the end of the twentieth century, under circumstances reminiscent of an Eric Ambler novel.

Early in the sixth century only three of the many works of Archimedes were generally known, those that appeared in the collection edited by Eutocius of Ascalon: *On the Equilibrium of Planes, On the Sphere and the Cylinder,* and the incomplete *On the Measurement of the Circle.* In the ninth century, Leo the Mathematician added to these the works *On Conoids and Spheroids, On Spirals, On the Quadrature of the Parabola,* the *Book of Lemmas,* and *The Sand Reckoner.* Leo's collection, known as Codex A, thus contained all of the works of Archimedes' in Greek now known, except *On Floating Bodies, On the Method, Stomachion,* and the *Cattle Problem.* This was one of two manuscripts available to William of Moerbeke when he made his translations of Archimedes in 1269. The other, known as Codex B, also called the Codex Mechanicorum, which contained only the mechanical works—*On the Equilibrium of Planes, On the Quadrature of the Parabola,* and *On Floating Bodies* (and possibly *On Spirals*)—was last referred to in the early fourteenth century; it then disappeared. Thus, as Marshall Clagett remarks, Codex A "was the source, directly or indirectly, of all the Renaissance copies of Archimedes."

Clagett also notes that it seems unlikely that Arab mathematicians possessed any collection of the works of Archimedes as complete as Codex A. According to Clagett, the writings of Archimedes available to the Arabs consisted of the following works: *On the Sphere and the Cylinder,* in an early-ninth-century translation revised in turn by Ishaq ibn Hunayn and Thabit ibn Qurra and reedited by Nasir ad-Din al-Tusi; *On the Measurement of the Circle,* translated by Thabit ibn Qurra and reedited by al-Tusi; a fragment of *On Floating Bodies;* possibly *On the Quadrature of the Parabola,* as evidenced by research on this work by Thabit ibn Qurra;

some indirect material of *On the Equilibrium of Planes*, as indicated in Greek mechanical works translated into Arabic; and other writings attributed to Archimedes by Arab mathematicians for which there is no extant Greek text, such as the *Book of Lemmas*, the *Book on the Division of the Circle into Seven Equal Parts*, and *On the Properties of the Right Triangle*.

Western Europe acquired its knowledge of Archimedes solely from Byzantium and Islam, for there is no trace of the earlier translations that Cassiodorus attributes to Boethius. The translation of Archimedean texts from the Arabic began in the twelfth century with *On the Measurement of the Circle*, a defective rendering that may have been done by Plato of Tivoli. A much superior translation of the same work was done by Gerard of Cremona, using an Arabic text due to Thabit ibn Qurra. The earliest-known Arabic translations of Archimedes are those of Thabit ibn Qurra. These comprise all the works of Archimedes' that have not been preserved in Greek, including the *Book of Lemmas*, *On Touching Circles*, and *On Triangles*.

The texts used by William of Moerbeke in his 1269 translations—Codices A and B—had come to the papal library in the Vatican from the collection of the Norman kings of the Two Sicilies. William translated all the works included in Codices A and B, except for *The Sand Reckoner* and Eutocius's *Commentary on the Measurement of the Circle*. William's translations did not include *On the Method*, the *Cattle Problem*, or the *Stomachion*, since these works were not in Codices A and B.

A new Latin translation of the works of Archimedes was done circa 1450 by James of Cremona, sponsored by Pope Nicholas V. James worked entirely from Manuscript A, and so his translation did not include *On Floating Bodies*, but it did have the two works in Codex A omitted by William of Moerbeke, *The Sand Reckoner* and Eutocius's *Commentary on the Measurement of the Circle*. Soon after James completed his translation the pope sent a copy to Nicholas of Cusa, who made use of it in his *De Mathematicis Complementis*, written in 1453–54. There are at least nine extant copies of this translation, one of which was corrected by Regiomontanus.

Codex A itself was copied several times, one copy being made by Cardinal Bessarion in the period 1449–68, and another by the humanist Georgio Valla, who used it in his *Outline of Knowledge*, printed at Venice in 1501. Copernicus, as we have learned, had a copy of the *Outline of Knowl-*

edge, in which he would have read Archimedes' account in *The Sand Reckoner* of the heliocentric theory proposed by Aristarchus of Samos, which preceded the Copernican theory by eighteen centuries.

Interest in Archimedes intensified from the mid-sixteenth century onward, and his influence can be seen in the works of Commandino, Simon Stevin, Kepler, Galileo, Torricelli, Leibniz, Newton, and many others. Translations were made into Italian, French, and German, and a new Latin edition was published in London in 1675 by Isaac Barrow, Newton's predecessor as Lucasian professor of geometry at Cambridge. At the end of the eighteenth century a new edition of the Greek text with Latin translation was prepared by the Italian mathematician Joseph Torelli (1721–81), published at Oxford after his death by Abram Robertson.

Nevertheless, a number of Archimedean writings remained missing, most notably the work *On the Method,* the existence of which was known only from references by Hero of Alexandria and the tenth-century Byzantine writer Suidas, who says that Theodosius of Bithynia wrote a commentary on it, though that was also lost.

The manuscripts Heiberg discovered were part of a palimpsest, in this case a euchologion, or prayer book, made up from recycled parchment leaves whose original contents had been scraped away and then written over with the new liturgical document. Heiberg's attention had been drawn to the euchologion through a report published in 1899 by the Greek scholar A. Papadopoulos-Kerameus, a catalog description of a manuscript collection in Istanbul belonging to the Metochion of the Holy Sepulchre, the daughter house of a famous monastery in Jerusalem. Papadopoulos-Kerameus had noted that the underlying script of the palimpsest MS 355 included a mathematical text, of which he printed a few lines in his catalog. Heiberg, who at the time was revising his edition of Archimedes, recognized the lines as being from an Archimedean work. He went to Istanbul and examined the palimpsest, first in 1906 and then again two years later, when he photographed the manuscript using the newly invented ultraviolet lamp. He reported on his discovery in 1907 in a long article in the scholarly journal *Hermes;* and in 1910–15 he incorporated his findings in the second edition of his three-volume opus on the works of Archimedes, upon which all subsequent Archimedean studies have been based. Meanwhile T. L. Heath translated *On the Method* into English, including it as a supplement to a new edition of his book *The Works of Archimedes,* published in 1912, which I used when I first began studying Archimedes.

Papadopoulos-Kerameus noted that the palimpsest contained a sixteenth-century inscription recording that it belonged to the ancient Palestinian monastery of St. Savas, known in Arabic as Mar Saba, founded in 483 a few miles east of Bethlehem on the west bank of the Jordan. The monastery had a renowned scriptorium for the copying and preservation of ancient manuscripts, of which its collection included more than a thousand works. Mar Saba was in ruins in 1625 when it was purchased by the Greek Orthodox Patriarchate of Jerusalem, which began a restoration in 1688. It has been suggested that in the early nineteenth century the euchologion and other ancient manuscripts in Mar Saba were taken for safekeeping to Istanbul; there they were preserved in the Jerusalem patriarchate's Metochion of the Holy Sepulchre, which was under the jurisdiction of the Patriarchate of Constantinople on the shore of the Golden Horn in the Fener quarter.

The German biblical scholar Constantine Tischendorf visited the Metochion in the early 1840s. He described the Metochion in his *Reise in den Orient* (Leipzig, 1846), where he says he found nothing of particular interest except a palimpsest whose pages included some mathematics. He appears to have stolen a page from the euchologion, for in 1879 a leaf from the palimpsest was sold from his estate to the Cambridge University Library. Nigel Wilson, a professor at Lincoln College, Oxford, examined this leaf in 1971 and identified it as being part of what by then had come to be known as the Archimedes Palimpsest.

The palimpsest disappeared from the Metochion not long after Heiberg's discovery, probably stolen in the chaos surrounding the fall of the Ottoman Empire and the creation of the new Republic of Turkey in 1923. Early in the 1920s the palimpsest was acquired by Marie Louis Sirieix, a French businessman and civil servant. In 1946 Sirieix gave the palimpsest as a wedding present to his daughter Anne Guersan, who had the euchologion restored and, as it appears, "embellished," by the addition of what proved to be forged images of the four Evangelists. Nigel Wilson stated that the images were "a disastrously misguided attempt to embellish the manuscript, presumably to enhance its value in the eyes of a prospective purchaser." In any event, the Guersan family put the palimpsest up for sale, and on 29 October 1998 it was auctioned at Christie's in New York, where it was purchased for $2 million by an anonymous buyer. The Jerusalem Patriarchate contested the auction in a lawsuit in New York, but the court ruled that the sale was legal.

Meanwhile, the anonymous buyer deposited the palimpsest with the Walters Art Museum in Baltimore, Maryland, in January 1999, providing funds for conservation, imaging, and scholarly study of the manuscript. A team of scientists from the Rochester Institute of Technology and Johns Hopkins University used computer processing of digital images of the underlying text of the palimpsest photographed in ultraviolet, infrared, and visible light. In May 2005 the palimpsest was irradiated with highly focused X-rays produced at the Stanford University Linear Accelerator Center in Menlo Park, California, which made it possible to read parts of the underlying text that had previously been undecipherable.

In 2002, John Lowden, a professor at the Courtauld Institute in London, deciphered a colophon on the palimpsest giving the date 13 April 1229, when the euchologion was dedicated after the recycling of the ancient manuscripts on which it was written. The palimpsest, now referred to as Codex C, contains parts of seven treatises by Archimedes as well as pages of four other works, including those of the fourth-century B.C. Attic orator Hyperides. The Archimedean writings include an almost complete text of the previously unknown work *On the Method;* a substantial part of *On Floating Bodies,* whose original Greek text had been lost; a page from the *Stomachion,* another unknown work of Archimedes'; and fragments of *On the Sphere and the Cyclinder, On Spirals, On the Measurement of the Circle,* and *On the Equilibrium of Planes.* Thus Codex C overlaps with Codices A and B for several works: together with A, it has a text of *On Spirals, On the Sphere and the Cylinder,* and *On the Measurement of the Circle;* along with B, it has a text of *On Floating Bodies.* Studies have shown that the Archimedean texts in the palimpsest, Codex C, were written in the second half of the tenth century, almost certainly in Constantinople.

The most important of these works by far is *On the Method,* whose full title is *On the Method Treating of Mechanical Problems, Dedicated to Eratosthenes.* Addressing Eratosthenes, the head of the Library of Alexandria, Archimedes explains the method by which he arrived at the propositions from which he deduced his theorems: "I thought fit to write out for you and explain in detail in the same book the peculiarity of a certain method, by which it will be possible for you to get a start to investigate some of the problems in mathematics by means of mechanics." The mechanical method used by Archimedes was one in which he mathematically balanced geometrical figures as if they were weights on a scale, comparing a figure of unknown area with one whose area was already known. Then, using the law of the lever, he determined the unknown area from the area that was known. He then extended his method to three dimensions so as to determine volumes, as he shows in Proposition 2 of the *On the Method:* "The cylinder with base equal to a great circle of the sphere and height equal to the diameter is ⅔ times the sphere [in area]." This was Archimedes' favorite theorem; he had it represented in a relief carved on his tombstone. Cicero found this monument in 75 B.C., when he was *quaester* of Sicily and the relief was still visible.

Archimedes used his mechanical method to determine the volumes of three solids of revolution—the ellipsoid, the paraboloid, and the hyperboloid—as well as the centers of gravity of the paraboloid and the hemisphere. He concluded the thesis by finding the volume of two solids, first the wedge cut from a right circular cylinder by two planes, and second the volume common to two equal right cylinders intersecting at right angles. All of these were results first achieved in western Europe only after the invention of the calculus by Newton and Leibniz, more than nineteen centuries after Archimedes had done the same in his lost work *On the Method.*

The treatise *On Floating Bodies,* which contains Archimedes' famous principal of buoyancy, the basis of hydrostatics, was previously known only from the Latin translation done in 1269 by William of Moerbeke, the Greek original having been lost. The text of this thesis in the Archimedes Palimpsest has considerable lacunae, so that William's translation is still used to give the undecipherable and missing parts of the Greek text.

Only a single page of the *Stomachion* was used in the palimpsest, the first page of the thesis, which became the last page in the euchologion. Otherwise, the only source for this thesis is a brief passage in an Arabic text published in Berlin in 1899, said to derive from a work by

Archimedes entitled *Stomachion*. Together the two sources are not enough to enable one to understand Archimedes' motive in writing this thesis, which appears to be about a type of geometrical game. The name *Stomachion* means literally "that which relates to the stomach," and it has been suggested that the game was so called because its difficulties were so great that it could give one a bellyache. E. J. Dijksterhuis, in his definitive book on Archimedes, concluded from his study of the ancient sources that the *Stomachion* "is a kind of game, played with bits of ivory in the form of simple planimetrical figures, the object being to fit these bits together in such a way that the various shapes of human beings, animals or different objects were imitated." He notes that the game board was apparently known to the Romans as the *loculus Archimedius*, or Archimedian box, and that "it consisted of fourteen bits of ivory of different forms, that these bits together formed a square, that it was possible to compose from them all sorts of figures (a ship, a sword, a tree, a helmet, a dagger, a column), and that this game was considered very instructive for children, because it strengthened the memory."

The single page of the *Stomachion* preserved in the palimpsest indicates that Archimedes was the first author in a field of mathematics now known as combinatorics, in this case finding the number of possible ways of fitting together the pieces on the game board. A recent analysis has shown that if there are 14 pieces on the board then there are 17,152 ways of arranging them. It is not known whether Archimedes found this solution, but anyone familiar with his work would bet that he did.

The Archimedes Palimpsest presently comprises 174 pages, three less than when Heiberg examined it, the missing pages probably removed when the euchologion was stolen from the Metochion. All but fifteen pages of the underlying text of the palimpsest have been deciphered, and these are now being analyzed at the Stanford Linear Accelerator Center. It takes about twelve hours to scan one page using an X-ray beam about the width of a human hair. Once each new page is analyzed it is posted online for the general public to examine.

After I viewed the most recent page posted on the Internet I went to visit the church of Agios Giorgios Metochi again, for I had not been there since before the 1998 rediscovery of the Archimedes Palimpsest. First I had to go to the office of the Jerusalem Patriarch in the headquarters of the Ecumenical Patriarch of Constantinople; after finding my way through a labyrinth of Byzantine bureaucracy, I finally received a document allowing me to visit the Metochion.

The church is within an extensive walled enclosure on a hillside above the Golden Horn, completely isolated from the tumultuous city around it, its main entryway closed by an enormous iron-barred wooden gate. I rang the bell several times, and when no one answered I picked up a stone and hammered on the door until finally the gate creaked open and a white-bearded old priest stuck his head out. I addressed him in Greek and showed him the document I had obtained from the patriarchate, whereupon he let me in and went to find the keys to the church.

The church itself had been restored since I'd last seen it, but everything else within the enclosure was in utter ruins, with a flock of goats grazing among the fallen columns and the other architectural fragments of the buildings that had once stood there. This was once the site of the palace of the Cantacuzenos family, Greeks from the Fener who ruled as hospodars of the trans-Danubian principalities of Moldavia and Wallachia under the aegis of the Ottoman sultan. All that remained of the palace was the shell of its chapel, dedicated to the Virgin, which had been the site of the Patriarchate of Constantinople in the late sixteenth century.

After seeing the church I sat for a while in the courtyard with the priest, who told me that he had been born in Istanbul but that after joining the priesthood he had been sent to the Monastery of the Holy Sepulchre in Jerusalem. He had returned only recently, he said, and was now living in virtual retirement in the Patriarchate of Constantinople, his only duty being to look after the church of Agios Giorgios Metochi. The church itself was all that was left of the Metochion of the Holy Sepulchre, the ruins of its other buildings indistinguishable from those of the Palace of the Hospodars of Moldavia and Wallachia. The survival of the church itself was an unexplained miracle, as was the preservation of its collection of ancient manuscripts until the early 1920s, when those that had not been stolen were removed to Athens for safekeeping.

The priest was not a learned man, and he knew nothing of the ancient manuscripts that had been preserved here. I showed him the copy of the latest page of the Archimedes Palimpsest that I had recently downloaded, and he was able to read some of it, though he had no idea what it meant since it was all in mathematical language. I told him that the works of Archimedes were written in the third century before Christ, and that a thousand years ago those that were in this palimpsest were copied by scribes in Constantinople, after which they had been preserved at the Mar Saba monastery in Palestine before being brought here to the Metochion early in the nineteenth century. The priest knew Mar Saba well, since he had been there several times during his years in Jerusalem. I then told him of the rediscovery of the manuscripts in the past century, and of the recovery of their underlying text at research laboratories in the United States. The priest shook his head in wonder after he heard my story, and he said that it must have been the will of God that these masterpieces of ancient science should have been preserved. I nodded as if I agreed with him, and then I thought of Archimedes himself, recalling the words with which he begins his treatise *On the Method:*

> Archimedes to Eratosthenes greeting.
>
> I sent you on a former occasion some of the theorems discovered by me, merely writing out the enunciations and inviting you to discover the proofs, which at the moment I do not give. The enunciation of the theorems which I sent were as follows.

And so now we have the proofs of Archimedes' theorems, lost for more than two thousand years, uncovered from a palimpsest that connects his time with ours through the intervening layers of the medieval Byzantine, Islamic, and Latin worlds.

Greetings, Archimedes; we await the next page from your manuscript.

HARRAN: THE ROAD TO
BAGHDAD

During the first week of June 2006, Bosphorus University hosted a conference together with the University of Harran, founded in 1993 in Urfa, a city in southeastern Turkey. Twenty miles to the southeast of Urfa is the ancient city of Harran, which in the mid-eighth century was briefly the capital of the eastern Islamic empire under Marwan II (r. 744–50), the last Umayyad caliph. A celebrated Islamic school of higher studies was founded there during the Abbasid caliphate, and it was this institution that gave its name to the modern University of Harran.

Urfa is on the northern edge of the great Mesopotamian plain, and Arabic is spoken there as well as Turkish, Kurdish, and even a few whispers of Syriac, the language of the tiny Christian community. The villagers in Harran are all Arabic speakers, though these days they also speak Turkish. The main road south from Urfa leads past Harran and through Syria and Iraq to Baghdad, following the ancient caravan route from central Anatolia to the confluence of the Euphrates and the Tigris.

Urfa has been identified as the biblical "Ur of the Chaldees," which Alexander the Great refounded as Edessa, after the city near his capital in Macedonia. Edessa became a famous city of religion and scholarship during the early centuries of the Byzantine period, and it played an important role both in the spread of Christianity and in the survival and transmission of ancient Greek learning.

During the medieval era Edessa was held in turn by the Byzantines, Sassanid Persians, Marwanid Kurds, Ayyubid Arabs, Armenians, Crusaders, and Seljuk Turks. An army of the First Crusade under Baldwin of Boulogne took Edessa from the Armenians in 1098. The city then became the capital of the county of Edessa, the first of several Crusader states founded in the Middle East. This state lasted less than half a century, for on Christmas Eve 1144 Edessa fell to the Seljuk Turks, who put all the Christian men to the sword and sold the women and children into slavery. The fall of Edessa made a deep impression on the West, leading Pope Eugenius III to proclaim the Second Crusade. The pope refers to Edessa in a letter he wrote on 1 December 1145 to King Louis VII of France, calling it "the city that . . . was ruled by Christians and alone served the Lord when, long ago, the whole world in the East was under the sway of pagans."

The pagans referred to by Eugenius were the Sabeans, who practiced an astral religion that originated in ancient Mesopotamia. The Sabeans prayed to spiritual beings whom they thought were the intermediaries between mankind and a Supreme Being, and who were the intelligences that guided the motion of the celestial bodies. Thus the Sabeans built temples to the sun, the moon, and the five planets, one of which has been excavated at a site called Sumatar Harabesi, some thirty-five miles east-southeast of Urfa.

The Second Crusade never succeeded in regaining Edessa, and the city passed in turn to the Ayyubid Arabs, Seljuk Turks, Mongols, Mamluks, and finally the Ottoman Turks, who took it in 1637. Under Muslim rule the city was known as Urfa, a variant of its ancient name, though the tiny Christian community continued to call it Edessa.

The first week of the conference was held in Istanbul on the campus of Bosphorus University, and for the second week it met on the campus of Harran University in Urfa. At the conclusion of the conference we spent a day in Harran, which I had visited for the first time twenty years before. The fragmentary ruins of the medieval Islamic city are now partly occupied by a village of mud-brick houses, each of them roofed with several conical domes, giving them the appearance of giant beehives; the only signs of the modern world are the television antennas and satellite dishes on the rooftops of what are probably the oldest dwellings in Turkey.

While we were in Urfa I presented a paper on the transmission of

Greek culture to Islam, during the first phase of which the Christian University of Edessa played a vital role. Works of Greek culture had been translated from Greek into Aramaic at the University of Edessa before it was closed in 489, after which some Nestorian scholars moved to the medical school at Jundishapur. Jundishapur had also received a group of Greek scholars after the Platonic Academy in Athens was closed in 529, among them Isidorus of Miletus, who would later return to Constantinople to help build the great church of Haghia Sophia. The Nestorian and Greek scholars at Jundishapur laid the foundations for the establishment of Greek culture in the medical school there, whose faculty would subsequently move to the House of Wisdom in Baghdad and translate these works into Arabic. It intrigued me that this transmission of culture included a Greek physicist from Miletus, where the first philosophers of nature had brought forth their ideas in the sixth century B.C., starting science on the long journey that eventually brought it through the Islamic world to western Europe and subsequently spread it worldwide.

Some of the most important work carried out in the House of Wisdom was done by Thabit ibn Qurra of Harran, one of the greatest scientists in the medieval Islamic world. Thabit, as we have learned, was a Syriac-speaking Sabean who earned his living as a money changer in Harran before he was discovered there by one of the Banu Musa and went to work as a translator at the House of Wisdom in Baghdad. There he translated from Greek and Syriac into Arabic the works of Aristotle, Archimedes, Euclid, Apollonius, Hero, Ptolemy, Nichomachus, Hippocrates, and Galen. Thabit's own work in mathematics, physics, astronomy, and medicine, translated from Arabic to Latin, was highly influential in the early development of European science. Roger Bacon refers to him as "the supreme philosopher among all Christians, who has added in many respects, speculative as well as practical, to the work of Ptolemy." But Thabit, as we know, was not a Christian; nor did he ever convert to Islam, for till the end of his days he remained a Sabean, which in Bacon's eyes would have made him a pagan or a heathen, a worshipper of the celestial bodies.

The Sabeans of Edessa and Harran would have inherited their knowledge of astronomy from sources originating in ancient Mesopotamia, part of the same tradition that gave them their astral religion. Apparently the Sabeans, like the Christians in southeastern Anatolia and

Mesopotamia, acquired their knowledge of Greek culture from the Alexandrian philosophical schools in late antiquity. In the case of the Sabeans this included not only rational science and philosophy but also the occult sciences of astrology, alchemy, magic, and hermeticism, or secret knowledge, named for the legendary Hermes Trismegistus.

Thabit ibn Qurra was privy to all of these traditions, inheriting the astronomical lore of Mesopotamia and both rational and occult science from Hellenistic Alexandria, and transmitting them to the Islamic world in Baghdad through his translations as well as his own writings. *MASI* lists 80 of his extant manuscripts, comprising 30 in astronomy, 29 in mathematics, 4 in history, 3 in mechanics, 3 in descriptive geography, 2 in philosophy, 2 in medicine, 2 in mineralogy, 2 in music, 1 in physics, 1 in zoology, and 1 in mysticism. Thabit's treatises on astronomy and mathematics were extremely influential in both Arabic and medieval European science, although his famous trepidation theory proved to be erroneous.

Thabit's four extant historical works are all concerned with the Sabeans, their history, chronology, religion, and customs. One of them, a Syriac manuscript entitled *Book of Confirmation of the Faith of Heathens*, proudly presents Thabit's claim that the Sabeans were heirs of the ancient pagan culture that civilized the world.

> We are the heirs and offspring of paganism which spread gloriously over the world. Happy is he who for the sake of paganism bears his burden without growing weary. Who has civilized the world and built its cities, but the chieftains and kings of paganism? Who made the ports and dug the canals? The glorious pagans founded all these things. It is they who discovered the art of healing souls, and they too made known the art of healing the body and filled the world with civil institutions and with wisdom which is the greatest of goods. Without them the world would be empty and plunged in poverty.

Thabit's single extant work on mysticism is entitled *Kitab al-Hiyal* (Book on Ingenious Manners). It survives only in medieval Latin translations bearing the titles *De Prestigious* (On Magic) or *De Imaginibus* (On

Images), which MASI describes as a "handbook for manufacturing metallic, wax and clay images of people, animals, cities or countries for magic operations connected with astrology." Recent research indicates that one of Thabit's grandsons who worked in Baghdad in the mid-tenth century taught his students the techniques of talismanic magic, which he presumably learned from Thabit. Two of the students of Thabit's grandsons were grandsons of the famous Andalusian physician al-Harrani, who practiced medicine at the court of the *amir* 'Abd al-Rahman II (r. 822–52) in Cordoba, and it was through them that Thabit's work on the occult sciences was introduced to Spain and subsequently to the Christian West.

My students are always surprised to find that Thabit would write a book on talismanic magic, which one of them described as "voodoo." I told her that scientists from the time of the ancient Egyptians, Babylonians, and Greeks had been involved in astrology, alchemy, divination, and magic. Thabit was working in Baghdad around the time when the *Thousand Nights and One Night* was written, when the glorious reign of Harun al-Rashid was still fresh in memory. He was the most famous scientist of his time, renowned for his knowledge of magic as well as his mathematics, and could have been a model for the villainous Moor—a student of "sorcery and spells, geomancy and alchemy, astrology, fumigation and enchantment"—who led Aladdin to his magical lamp.

Thabit stands in a long line of wizard-scientists stretching from Pythagoras to Newton, for in the popular mind those who achieve an understanding of nature gain power over it. Most modern historians focus only on the rational side of the developments that led to the Scientific Revolution, leaving out of consideration what Plato called "the spindle of Necessity on which all the revolutions turn," referring to the heavenly spheres that once carried the celestial orbs in divinely ordained harmony.

The ancient Sabean faith continues to the present across the border from Harran in Syria, as well as in Iraq, as I learned only a few years ago. Harran and other beehive villages to its south in Syria are inhabited by seminomadic Arabs who herd their flocks in the Jullab plain, the northernmost extension of the great Mesopotamian desert. Until two decades ago, when I first visited Harran, the way of life of the villagers had not changed since biblical times, with desert nomads gradually abandoning their wandering ways and settling down for a time here on the northern margin of the Mesopotamian plain before moving on again, just as the family of Abraham did when they passed this way three millennia ago. Genesis 12:1–5 tells of how Abraham left Harran in obedience to the call of Yahweh:

> Yahweh said to Abram, "Leave your country, your family and your father's house, for the land I will show you. I will make you a great nation; I will bless you and make your name so famous that it will be used as a blessing.... So Abram went as Yahweh told him, and Lot went with him. Abram was seventy-five years old when he left Haran. Abram took his wife Sarai, his nephew Lot, all the possessions they had amassed and the people they had acquired in Haran. They set off for the land of Canaan and arrived there.

Twenty miles south of Harran the main road crosses the border into Syria, where there are about a dozen villages inhabited by Arabs who still hold to the ancient Sabean faith. The road then continues south for another fifty miles to Al-Raqqa, an ancient crossroads on the left bank of the Euphrates.

Al-Raqqa was refounded in 771 by the caliph al-Mansur. Harun al-Rashid built a palace north of Al-Raqqa in the closing years of the eighth century, calling it Kasr as-Salam, the Palace of Peace. Thenceforth this became his favorite residence, and Al-Raqqa became the second capital of the Abbasid caliphate after Baghdad.

The astronomer al-Battani, a younger contemporary of Thabit ibn Qurra's from Harran, built an observatory in Al-Raqqa in the last quarter of the ninth century. The astronomical tables he compiled there were translated into Latin and were used by both Copernicus and Tycho Brahe. Al-Battani was also a Sabean, but, unlike Thabit ibn Qurra, he converted to Islam. Nevertheless, he remained proud of his Sabean heritage, as evidenced by the title of his great work on astronomy, *Zij al-Sabi*, the Sabean Tables.

Just to the south of Al-Raqqa the road from Urfa crosses the Euphrates and joins the main highway from Aleppo to Baghdad. Beyond the junction the highway runs along the right bank of the Euphrates, following the track of the oldest caravan route in the world, which has brought goods and ideas back and forth between Mesopotamia and the Mediterranean since the beginning of recorded history and even before. Thabit ibn Qurra would have followed this route on his way to Baghdad, bringing with him the knowledge of Greek and ancient Mesopotamian astronomy that he translated into Arabic in the House of Wisdom.

Thabit arrived in Baghdad circa 862, a century after its founding as the Abbasid capital. A century later the geographer Muqaddasi could with justice describe the Abbasid capital as a city beyond compare, an encomium that I recalled in Harran, thinking of Thabit ibn Qurra setting out on the road to Baghdad:

> Baghdad, in the heart of Islam, is the city of well-being; in it are the talents of which men speak, and elegance and courtesy. Its winds are balmy and its science penetrating. In it are to be found the best of everything and all that is beautiful. From it comes everything of consideration, and every elegance is drawn towards it. All hearts belong to it, and all wars are against it.

And that brought to mind lines by the poet Khuraymi, lamenting the destruction of Baghdad in a civil war in 812–13. This threnody has a particular resonance now, when the city is once again being ravaged by internal strife:

> *Behold Baghdad! There bewildered sparrows*
> *Build no nests in its house.*
> *Behold it surrounded by destruction, encircled*
> *With humiliation, its proud men besieged.*

But Baghdad was restored and flourished for another four centuries before it was destroyed again, and during that time Thabit ibn Qurra and those who succeeded him in the House of Wisdom continued their translations and researches, which in time were passed on to the West.

Such is the story of how Greek science came to western Europe through the lands of Islam, passing through Harran on the road to Baghdad, where I ended an intellectual quest that began in Miletus, led along the way by the power of Aladdin's lamp.

ACKNOWLEDGMENTS

I would like to acknowledge the generous assistance that I have received from the librarians at the American Research Institute in Turkey and Boğaziçi University in Istanbul. I would particularly like to thank Dr. Anthony Greenwood, director of the American Research Institute in Turkey; Professor Taha Parla, director of the Bosphorus University Library; and Hatice Ün, the assistant director. I would also like to thank my editor, Ann Close, who has been of great help to me in giving my manuscript its final form.

ILLUSTRATION CREDITS

NOTES

ABBREVIATIONS

ARI: Alexandria, Real and Imagined, ed. Anthony Hirst and Michael Silk
DSB: Dictionary of Scientific Biography, 16 vols., ed. Charles Coulston Gillespie
EGS: Early Greek Science: Thales to Aristotle, G. E. R. Lloyd
EHAS: Encyclopedia of the History of Arabic Science, 3 vols., ed. Roshdi Rashed and Regis
   Morelon
GPTP: Greek Philosophy, Thales to Plato, John Burnet
GSAA: Greek Science After Aristotle, G. E. R. Lloyd
*GTAC: Greek Thought, Arabic Culture: The Graeco-Arabic Translation Movement in Baghdad
   and Early Abbasid Society*, Dimitri Gutas
HAA: A History of Arabic Astronomy, George Saliba
HGP: A History of Greek Philosophy, 6 vols., William K. C. Guthrie
HMES: A History of Magic and Experimental Science, 8 vols., Lynn Thorndike
LA: The Library of Alexandria, ed. Roy MacLeod
LEP: Lives of Eminent Philosophers, Diogenes Laertius
LMS: The Legacy of Muslim Spain, 2 vols., ed. Salma Khadra Jayyusi
*MASI: Mathematicians, Astronomers and Other Scholars of Islamic Civilization and Their
   Works (7th–14th Centuries)*, B. A. Rozenfeld and Ekmeleddin Ihsanoğlu
MEMS: Medieval and Early Modern Science, 2 vols., A. C. Crombie
MPNP: Mathematical Principles of Natural Philosophy, Isaac Newton
NAR: Never at Rest: A Biography of Isaac Newton, Richard S. Westfall
ORHS: On the Revolutions of the Heavenly Spheres, Nicholas Copernicus
PL: Plutarch's Lives, trans. Bernadotte Perrin
PWD: The Philosophical Writings of Descartes, trans. John Cottingham, et al.
RGOES: Robert Grosseteste and the Origins of Experimental Science, A. C. Crombie
SCI: Science and Civilization in Islam, Seyyed Hossein Nasr
SHMS: Studies in the History of Mediaeval Science, Charles Homer Haskins
SMMA: The Science of Mechanics in the Middle Ages, Marshall Clagett

STLOE: *Science, Technology and Learning in the Ottoman Empire*, Ekmeleddin Ihsanoğlu
TCT: *Three Copernican Treatises*, Edward Rosen
TPP: *The Presocratic Philosophers*, G. S. Kirk and J. E. Raven

INTRODUCTION

4 "the Graeco-Arabic translation movement": Gutas, *GTAC*, p. 8.
4 As Edward Said remarked: Said, *Culture and Imperialism*, p. xxv.

1 IONIA: THE FIRST PHYSICISTS

5 "the glory of Ionia": Herodotus, *Histories*, book V, p. 28.
5 "many are the achievements of this city": Strabo, *Geography*, vol. 6, pp. 206–7.
6 "had the good fortune": Herodotus, *Histories*, book I, p. 142.
6 "The Ionian countryside": Pausanias, *Description of Greece*, vol. 1, p. 240.
6 "the wonders of Ionia are numerous": Ibid., p. 245.
6 Yet in Delos: Hesiod, *the Homeric Hymns and Homerica*, vol. 3, pp. 148–58.
8 "the Egyptians": Herodotus, *Histories*, book II, p. 4.
8 "knowledge of the sundial": Ibid., book II, p. 109.
8 "fairest of all the stars": Edmonds, *Lyra Graeca*, vol. 1, p. 203.
9 "first founder": Guthrie, *HGP*, vol. 1, p. 40.
10 predicted the total eclipse: Herodotus, *Histories*, Book I, p. 74.
10 "Thales, who led the way": Guthrie, *HGP*, vol. 1, p. 55.
10 "from the observation": Ibid., vol. 1, p. 55.
10 "the first of the Greeks": Ibid., vol. 1, p. 72.
10 "solstices, times, seasons and equinoxes": Ibid., I, p. 74.
11 "separated off": L. Taran, "Anaximander," *DSB*, 1, pp. 150–51.
11 "The tale is not mine": Guthrie, *HGP*, vol. 1, p. 69.
11 "hangs freely": Ibid., p. 98.
11 "He says moreover": Ibid., p. 102.
11 "all things proceed from one": Ibid., p. 115.
12 "differs in rarity": Ibid., p. 121.
12 "like a leaf": Burnet, *GPTP*, p. 20.
12 "As our soul": Ibid., p. 19.
12 "The Lord [Apollo]": Kirk and Raven, *TPP*, p. 211.
12 "What I understood": Guthrie, *HGP*, vol. 1, p. 412.
12 "*Panta rhei*": Burnet, *GPTP*, p. 46.
12 "Heraclitus somewhere says": Kirk and Raven, *TPP*, p. 197.
13 "God is day night": Ibid., p. 191.
13 "Evil witnesses": Ibid., p. 189.
14 "magical arts and Pythagorean numbers": Thorndike, *HMES*, vol. 1, p. 370.
14 "The wise men": Guthrie, *HGP*, vol. 1, p. 209.
15 "They supposed": Kirk and Raven, *TPP*, p. 237.
15 "They said that the bodies": Cohen and Drabkin, *Source Book in Greek Science*, p. 96.

15 "They account for it": Guthrie, *HGP*, vol. 1, p. 296.
15 *There's not the smallest orb:* Shakespeare, *The Merchant of Venice*, act V, scene I.
16 "ascribed to the gods": Guthrie, *HGP*, vol. 1, p. 371.
16 "the Ethiopians say": Kirk and Raven, *TPP*, p. 168.
16 "God is one": Guthrie, *HGP*, vol. 1, p. 374.
16 "Stop, do not beat him": Lloyd, *EGS*, p. 11.
16 "let custom": Kirk and Raven, *TPP*, p. 271.
17 "Either a thing is": Ibid., p. 269.
17 *Then gin I thinke:* Spenser, *The Faerie Queene*, book 7, canto VIII, l. 2.
18 "But come,": Kirk and Raven, *TPP*, p. 325.
18 "roots of everything": Zeller, *Outline of the History of Greek Philosophy*, p. 56.
18 "from these things": Kirk and Raven, *TPP*, pp. 328–29.
18 "the elements are": Ibid., pp. 329–30.
18 "the groundwork bee": Spenser, *The Faerie Queene*, book 7, canto VII, l. 25.
19 "I, an immortal god": Lloyd, *EGS*, p. 138.
20 "Nothing occurs at random": Kirk and Raven, *TPP*, p. 413.
20 there are innumerable worlds: Guthrie, *HGP*, vol. 2, p. 405.
20 "We must suppose": Kirk and Raven, *TPP*, p. 378.
20 "the sun, the moon": Ibid., p. 391.
21 But the man: Plutarch, *PL*, "Pericles," IV, 4.
21 *And they learned:* Guthrie, *HGP*, vol. 1, p. 365.

2   CLASSICAL ATHENS: THE SCHOOL OF HELLAS

23 "Mighty indeed": Thucydides, *History of the Peloponnesian War*, vol. 2, p. 41.
23 "open to the world": Ibid., vol. 2, p. 39.
24 "from a waterless": Plutarch, *PL*, "Cimon," XIII, 8.
24 You will spend your time: Aristophanes, *The Clouds*, 1008ff.
25 "the olive grove of Academe": Milton, *Paradise Regained*, book 4, ll. 244ff.
25 "what we have in mind": Plato, *Laws*, I, 643e.
25 "the philosophers make": Guthrie, *HGP*, vol. 4, p. 21.
25 "but that they might": Ibid., vol. 4, p. 21.
25 "Zeno and Parmenides": Plato, *Parmenides*, 127b–c.
26 "This magnificent hope": Plato, *Phaedo*, 98c.
26 "only along the lines": Plato, *Timaeus*, 59d.
26 "everlasting and unwandering stars": Ibid., 40b.
27 "the spindle of Necessity": Plato, *Republic* X, 617c.
27 "we must require": Ibid., VII, 527c.
27 "true numbers" and "true motions": Ibid., VII, 529d.
27 "Let's study astronomy": Ibid., VII, 530b–c.
29 "on what hypotheses": Guthrie, *HGP*, vol. 5, p. 450.
32 "Aristotle that hath an oare": Montaigne, translated by Florio, quoted by Guthrie, *HGP*, vol. 6, p. ix.
32 "Now intelligent action": Aristotle, *Physics* vol. 2, 8: 12–15.
35 "Heraclides supposed": Lloyd, *EGS*, p. 95.
36 "a distinguished man": Diogenes Laertius, *LEP*, v. 58.

36 **"For if one observes"**: Simplicius, *Commentary on Aristotle's Physics*, 916, 10ff., quoted in Lloyd, *GSAA*, p. 16.
36 **"If we are not troubled with doubts"**: Furley, "Epicurus," *DSB*, vol. 4, p. 381.
37 **"swerve"**: Lucretius, *De Rerum Natura* vol. 1, p. 62.

3 HELLENISTIC ALEXANDRIA: THE MUSEUM AND THE LIBRARY

39 **"The Museum is also"**: Strabo, *Geography*, XVII, Part I, p. 8.
39 **"had at his disposal"**: Mostafa El-Abbadi, "The Alexandria Library in History," in Hirst and Silk, *ARI*, p. 171.
39 **"the beautiful city of Alexandria"**: Robert Barnes, "Cloistered Bookworms," in MacLeod, *LA*, p. 66.
40 **"the first library"**: Ibid., p. 66.
41 **"lived in the time"**: John Murdoch, "Euclid," *DSB*, 4, p. 414.
41 **"has exercised an influence"**: Ibid., p. 415.
41 **"collecting many of the theorems"**: Ibid., p. 423.
42 **"diversions of a geometry"**: *PL*, "Marcellus," XIV, 4.
42 **"it is not possible"**: Ibid., xvii.
43 **"Give me a place to stand"**: Simplicius, *Commentary on Aristotle's Physics*, quoted in Dijksterhuis, *Archimedes*, p. 16.
43 **"without a moment's delay"**: Vitruvius, *De Architectura*, vol. 9, p. 3.
44 **"a volume equal to that"**: Dijksterhuis, *Archimedes*, p. 362.
44 **"Aristarchus of Samos"**: Ibid., pp. 362–63.
45 **"on the ground that he was disturbing"**: Plutarch, "Concerning the Face Which Appears in the Orb of the Moon," *Moralia*, vol. 12, p. 923.
48 **"almost exclusively"**: A. G. Drachmann, "Hero of Alexandria," *DSB*, 6, p. 310.
48 **"Thus," he writes**: Cohen and Drabkin, *Source Book in Greek Science*, p. 250.
53 **"of things on land"**: Strabo, *Geography*, vol. 1, p. 3.
53 **"the home of men"**: Ibid., vol. 1, p. 443.
53 **"its inhabitants are more savage"**: Ibid., vol. 2, p. 259.
53–4 **"in the middle of the heavens"**: Ptolemy's *Almagest*, p. 41.
55 **"we shall dismiss"**: Lloyd, *GSAA*, pp. 130–31.
57 **"the quickening of the pulse"**: Thorndike, *HMES*, vol. 1, pp. 144–45.
57 **"God granted him"**: Cohen and Drabkin, *Source Book in Greek Science*, p. 27.
57 **"It is impossible to divide a cube"**: Heath, *Diophantus of Alexandria*, pp. 145–46.
57 **"which this margin"**: Boyer, *A History of Mathematics*, p. 354.
58 **"Bees . . . by virtue"**: Ibid., p. 199.
58 **"I therefore swear"**: Ivor Bulmer-Thomas, "Pappus of Alexandria," *DSB*, 10, p. 301.
58 **"philosophy and magic"**: Thorndike, *HMES*, vol. 1, p. 290.
59 **"On the inner side"**: Mostafa El-Abbadi, "The Alexandria Library in History," in Hirst and Silk, *ARI*, pp. 173–74.
59 **"certain ancient astrologers"**: Carmody, *The Astronomical Works of Thabit Ibn Qurra*, pp. 45–46.
60 **What then has Athens**: Lloyd, *GSAA*, p. 168.

4 FROM ATHENS TO ROME, CONSTANTINOPLE, AND JUNDISHAPUR

62 "he fled from the gloom": Morford, *The Roman Philosophers*, p. 25.
62 "Live unnoticed": David J. Furley, "Lucretius," *DSB*, 8, p. 536.
62 "this very superstition": Lucretius, *De Rerum Natura*, vol. 1, pp. 82–83.
62 "Nothing is ever produced": Ibid., vol. 1, pp. 150–51.
62 "time by itself does not exist": Ibid., vol. 1, pp. 459–61.
62 "What can be a surer guide": Ibid., vol. 1, 699–700.
63 "at quite indeterminate times and places": Ibid., vol. 2, pp. 218–20.
63 "that by perusing about 2,000 volumes": Pliny the Elder, *Natural History*, vol. 1, 17.
63 "a diffuse and learned work": David E. Eicholz, "Pliny," *DSB*, 11, p. 39.
64 "No one should wonder": Thorndike, *HMES*, vol. 1, pp. 59–60.
64 "counted the stars": Seneca, *Natural Questions*, VII, 25.
64 "It is enough for Christians": Clagett, *Greek Science in Antiquity*, pp. 132–33.
69 "a man eloquent and greatly skilled": Ibid., p. 181.
69 "their valuable methods": Boyer, *A History of Mathematics*, p. 238.
69 "Though I am a Hellene": Runciman, *The Last Byzantine Renaissance*, p. 22.
70 "incorporeal motive force": Lloyd, *GSAA*, p. 159.

5 BAGHDAD'S HOUSE OF WISDOM: GREEK INTO ARABIC

72 al-Mansur "was the first caliph": Gutas, *GTAC*, p. 30.
74 "The people of every age": Ibid., p. 46.
74 "mistress of all sciences": Ibid., p. 108.
75 "he translated from Persian": Ibid., p. 55.
75 "was employed full-time": Ibid., p. 55.
76 "encouraged me to compose": Ibid., p. 113.
77 "From his earliest youth": *Thousand Nights and One Night*, vol. 3, p. 382.
77 "At the beginning of his reign": Gutas, *GTAC*, pp. 77–78.
78 "The sages": Thorndike, *HMES*, vol. 1, p. 65.
79 "for full-time translation": Gutas, *GTAC*, p. 133.
79 "the land of the Greeks": G. C. Anawati, "Hunayn ibn Ishaq," *DSB*, 15, p. 230.
80 "I sought for it": Ibid., p. 230.
82 "a certain great noble": Thorndike, *HMES*, vol. 1, p. 655.
82 Did you not see: André Clot, *Harun al-Rashid*, p. 33.

6 THE ISLAMIC RENAISSANCE

83 "humorless, gross and dull": Turner, *Science in Medieval Islam*, p. 120.
84 "He it is who": *Koran*, x, 6.
84 "none in the heavens": Ibid., xxvii, 65.
84 "What can you know": Sa'di of Shiraz, *Tales from the Bustun . . . and Gulistan*, p. 26.
85 "The best gift from Allah": Turner, *Science in Medieval Islam*, p. 131.
86 *When shall it be*: Nasr, *SCI*, p. 206.

89 "decrees of the stars": Turner, *Science in Medieval Islam*, p. 109.
90 "the possible gravity": Nasr, *SCI*, p. 133.
90 "borrowed power": Crombie, *MEMS*, vol. 2, p. 53.
91 *Ah, but my Computations:* Khayyam, *The Rubaiyat*, lxii.
94 "Baghdad, which has no equal": Clot, *Harun al-Rashid*, p. 151.

7 CAIRO AND DAMASCUS

97 "good fortune": A. I. Sabra, "Ibn al-Haytham," *DSB*, 6, p. 190.
97 "doctrines whose matter": Ibid., p. 190.
98 "visual rays": Ibid., p. 192.
98 "distinct form": Ibid., p. 193.
99 "more truly descriptive": Ibid., p. 198.
100 "scientific intuition": Ibid., p. 203.
102 "Galen's medicine": Johnson, *A History of the Jews*, p. 186.
104 I therefore asked Almighty God: David A. King, "Ibn al-Shatir," *DSB*, 12, p. 358.

8 AL-ANDALUS, MOORISH SPAIN

107 "the bride of al-Andalus": Robert Hillenbrand, "The Ornament of the World," *LMS*, p. 118.
107 "the ornament of the world": Ibid., p. 118.
107 "In four things Cordoba": Ibid., p. 118.
109 "one of the books of the Christians,": Juan Vernet and Julio Samso, "Development of Arabic Science in Andalusia," *EHAS*, vol. 1, p. 246.
109 "Only by repeated visits": Sami Hamarneh, "al-Zahrawi," *DSB*, 14, p. 585.
110 "the bringer of joy": Ibid., p. 585.
110 "applied himself": Vernet and Samso, "Development of Arabic Science in Andalusia," *EHAS*, vol. 1, p. 254.
110 "very wise . . . philosopher": Thorndike, *HMES*, vol. 2, p. 813.
110 "a compendium of magic": Juan Vernet, "Al-Majriti," *DSB*, 6, pp. 39–40.
110 "confused compilation of extracts": Thorndike, *HMES*, vol. 2, p. 815.
111 "to explain what": Yvonne Dold-Samplonius and Heinrich Hermelink, "al-Jayyani," *DSB*, 7, p. 82.
112 *His tables Toletanes:* Chaucer, "The Franklin's Tale," *Canterbury Tales*, 545–56.
113 "oval" rather than circular: Juan Vernet, "Al-Zarqali," *DSB*, 14, p. 595.
113 *I've a sickness doctors can't cure:* Menocal, *The Ornament of the World*, p. 112.
113 "I have observed women": Menocal, et al., *The Literature of Al-Andalus*, p. 238.
114 "combined the songs": James T. Monroe, "Zajal and Muwashshaha: Hispano-Arabic Poetry and the Romance Tradition," *LMS*, p. 412.
117 "own object of desire": Davidson, *Alfarabi, Avicenna, and Averroes, on Intellect*, p. 223.
117 Our society allows no scope: Robert Hillenbrand, "The Ornament of the World," *LMS*, p. 122.
118 "the man whose theory shook": Julio Samso, "Al-Bitruji," *DSB*, 15, p. 35.

9 FROM TOLEDO TO PALERMO: ARABIC INTO LATIN

120 "These are the twenty-eight": Thorndike, *HMES*, vol. 1, p. 712.
121 "he learned what the song": Ibid., p. 705.
123 "for in Sicily and Salerno": Haskins, *SHMS*, pp. 132–33.
123 "all the secrets of philosophy": Ibid., p. 135.
124 "long period of study": Crombie, *MEMS*, vol. 1, p. 10.
124 "enunciations": "Adelard of Bath," *DSB*, 1, p. 63.
124 *editio specialis:* Ibid.
124 "simple ignorant": John Murdoch, "Euclid," *DSB*, 4, p. 449.
124 "For nother is there anie": Ibid., p. 449.
124 "manifolde additions": Ibid., p. 450.
125 "something new from my Arab studies": Crombie, *MEMS*, vol. 1, p. 10.
125 "better to attribute": Ibid., p. 26.
125 "the opinions of the Arabs": Haskins, *SHMS*, p. 41.
127 "John, son of David": Ibid., p. 13.
127 "this pearl of philosophy": Thorndike, *HMES*, vol. 2, p. 270.
128 "there, seeing the": Richard Lemay, "Gerard of Cremona," *DSB*, 15, p. 174.
128 "wiser philosophers of the world": Haskins, *SHMS*, p. 127.
128 "Gerard of Toledo": Richard Lemay, "Gerard of Cremona," *DSB*, 15, p. 173.
130 "secrets of all divinity": Thorndike, *HMES*, vol. 2, p. 216.
130 "In the name of God": Ibid., p. 182.
131 "If you live among the Muslims": S. Maqbul Ahmad, "Al-Idrisi," *DSB*, 7, p. 8.
131 "the wonder of the world": Kantorowicz, *Frederick the Second*, p. 356.
132 "baptized sultans": Haskins, *SHMS*, p. 243.
132 "classified in order": Masson, *Frederich II*, p. 224.
132 "We have followed Aristotle": Ibid., p. 216.
132 "I have heard from": Ibid., p. 112.
133 "with the new Indian numerals": Kurt Vogel, "Leonardo Fibonacci," *DSB*, 4, p. 604.
133 "How many pairs of rabbits": Boyer, *A History of Mathematics*, p. 287.
134 "the late Theodore": Haskins, *SHMS*, p. 247.
135 "And how is it": Ibid., p. 267.
135 "whether one soul": Ibid., p. 266.
135 "a notable inquirer": Thorndike, *HMES*, vol. 2, p. 314.
135 "That other, round the loins": Dante, *Inferno*, Canto XX.
136 *A wizard, of such:* Haskins, *SHMS*, p. 19.

10 PARIS AND OXFORD I: REINTERPRETING ARISTOTLE

138 "until they shall": Rashdall, *The Universities in Europe*, vol. 1, p. 357.
138 "double-truth": Crombie, *MEMS*, vol. 1, p. 64.
139 "composition" and "resolution": Crombie, *RGOES*, pp. 62–63.
139 "This method involves": Ibid., p. 63.
140 "the same cause": Ibid., p. 85.
140 "because nature operates": Ibid., p. 86.
140 "by experience and reason": Ibid., p. 87.

140 "the science that is concerned": Ibid., p. 91.
140 "reason for the fact": Ibid., p. 96.
140 "Metaphysics of Light": Ibid., pp. 128–34.
141 "when the sounding body": Ibid., p. 114.
141 "multiplication of species": Ibid., p. 109–10.
141 "untouched and unknown": Ibid., p. 119.
142 "to give solace to the earth": Lorris and Meun, *The Romance of the Rose*, 83, 109ff.
142 "These modes of celestial": Crombie, *RGOES*, p. 97.
144 "to make all parts of philosophy": Ibid., p. 19.
144 "In this sixth book": Crombie, *MEMS*, vol. 1, pp. 147–48.
144 "For the saints": Thorndike, *HMES*, vol. 2, p. 552.
144 "great accidents and": Ibid., p. 583.
144 "a man in every science": Ibid., p. 527.
144 "What wonder that a man": Ibid., p. 528.
145 "even though the natural": quoted in Lindberg, *The Beginnings of European Science*, p. 232.
146 "what remedies you think": Thorndike, *HMES*, vol. 2, p. 625.
147 "in the things of the world": Ibid., p. 144.
147 "Every multiplication": Ibid.
147 "three great prerogatives": Ibid., p. 141.
147 "no science can be known": Crombie, *RGOES*, p. 114.
148 Machines for navigation: Crombie, *MEMS*, vol. 1, p. 55.
148 "it has been proved": Thorndike, *HMES*, vol. 2, p. 655.
148 "inspires the intellect": Ibid., p. 657.
148 "after some noyse": Sandys, "Roger Bacon in English Literature," in Little, *Roger Bacon: Essays*, p. 362.

11 PARIS AND OXFORD II: THE EMERGENCE OF EUROPEAN SCIENCE

149 "science of weights": Edward Grant, "Jordanus de Nemore," *DSB*, 7, p. 172.
149 "positional gravity": Ibid., p. 172.
149 "weight is heavier positionally": Ibid.
150 "if the arms of the balance": Clagett, *SMMA*, p. 75.
150 "work," the product of the weight: Ibid., pp. 8, 78–79.
150 "virtual velocity": Ibid., pp. 8, 78–79, 83.
150 "axiom of Jordanus": Ibid., pp. 78–79.
152 *calculatores*, those who studied: Crombie, *MEMS*, vol. 2, p. 89.
152 mean speed rule: Ibid., p. 93.
152 "difform": Ibid., p. 90.
153 "on the supposition": Ernest A. Moody, "Jean Buridan," *DSB*, 2, p. 604.
153 "are not immediately evident": Ibid., pp. 604–5.
153 impetus theory: Ibid., p. 606.
154 "would endure forever": Ibid., p. 606.
154 "quantity of matter": Ibid.
154 immaterial "intelligences": Clagett, *SMMA*, p. 520.
154 "For it could be said": Ernest A. Moody, "Jean Buridan," *DSB*, 2, p. 606.

154 "indisputably true that": Ibid., p. 607.
155 Buridan had an affair: Ibid., p. 603.
156 "it is not impossible": Marshall Clagett, "Nicole Oresme," DSB, 10, p. 223.
156 "subject to correction": Crombie, MEMS, vol. 2, p. 78.
156 "For God fixed the earth": Ibid., p. 82.
157 "Completed in camp": Edward Grant, "Peter Peregrinus," DSB, 10, p. 533.
161 Forthwith / As clock, that calleth up: Dante, Paradiso, x.
162 "It would be futile": Crombie, RGOES, p. 216.
162 "of corporeal influences": Ibid., p. 214.
162 "there is something wonderful": Lorenzo Minio-Paluello, "William of Moerbeke," DSB, 9, p. 435.
163 "a globe of water": William A. Wallace, "Dietrich von Freiberg," DSB, 4, p. 93.

12 FROM BYZANTIUM TO ITALY: GREEK INTO LATIN

165 There were present: Haskins, SHMS, p. 144.
165 "James, a clerk of Venice": Ibid., p. 144.
166 "a man most learned": Ibid., p. 143.
167 "in spite of the hard work": Lorenzo Minio-Paluello, "William of Moerbeke," DSB, 9, p. 435.
167 "occult" inquiry: Ibid., p. 435.
168 "the regulative power": Thorndike, HMES, vol. 2, p. 886.
168 "the brain is the seat": Ibid.
168 "Is life possible": Ibid.
168 "perpetual and incorruptible": Ibid., p. 900.
168 "a man who appeared": Ibid., p. 889.
168 "the idle curiosity": Ibid., p. 890.
170 "towards the sea": A. A. Vasiliev, History of the Byzantine Empire, vol. II, p. 623.
170 "From morning to evening": Ibid., p. 702.
173 "the most original of all Byzantine thinkers": Runciman, The Last Byzantine Renaissance, p. 2.
173 "Neither, it will not": Nigel Wilson, From Byzantium to Italy, p. 56.
174 "If the immensity": Ibid.
174 "If you sail": Ibid.
177 "the rebirth of trigonometry": Boyer, A History of Mathematics, p. 308.

13 THE REVOLUTION OF THE HEAVENLY SPHERES

178 "God be praised": Pastor, The History of the Popes, vol. 6, p. 149.
178 "All the world is in Rome": Ibid., p. 149.
178 "this remote corner of the earth": Copernicus, ORHS, Preface, p. 5.
180 "not so much the pupil": Rosen, Narratio Prima, in Rosen, TCT, p. 111.
180 "after sunset on the seventh": Copernicus, ORHS, IV, 27, p. 249.
180 "lectured on mathematics": Rosen, Narratio Prima, in Rosen, TCT, p. 111.
180 "nineteenth year of Hadrian": Copernicus, ORHS, IV, 14, p. 224.
180 "to determine the positions": Ibid., p. 223.

181 "a manuscript of six leaves": Gassendi, *The Life of Copernicus*, p. 140.
181 "to the meridian of Cracow": Rosen, "Nicholas Copernicus," *DSB*, vol. 3, 402.
182 "the apparent motion": Rosen, *Commentariolus*, in Rosen, *TCT*, p. 57.
182 "in which everything": Ibid., pp. 57–58.
182 "imperceptible in comparison": Ibid., p. 58.
182 "the apparent retrograde": Ibid., p. 59.
182 "the motion of the earth": Ibid., p. 58.
182 The celestial spheres: Ibid., pp. 59–60.
183 "Then Mercury runs": Ibid., p. 90.
184 "my teacher": Rosen, *Narratio Prima*, in Rosen, *TCT*, p. 109.
184 "We look forward": Ibid., p. 122.
184 "by having the sun": Ibid., pp. 135–36.
185 "I know that he always": Armitage, *Sun, Stand Thou Still*, p. 126.
185 "You have in this recent": Gingerich, *The Book Nobody Read*, p. 20.
185 "He had lost his memory": Armitage, *Sun, stand thou still*, p. 127.
186 "The sun stood still": Joshua 10:12–14.
186 "People give ear": Kuhn, *The Copernican Revolution*, p. 191.
187 "I can reckon easily enough": Copernicus, *ORHS*, p. 2.
187 "I myself think": Ibid., pp. 19–20.
188 In the center of all: Ibid., pp. 25–26.
189 al-Tusi couple: Saliba, *HAA*, p. 246.
189 Though the courses: Thomas W. Africa, "Copernicus' Relation to Aristarchus and Pythagoras," p. 407.
190 "assuming that the heavens": Ibid., p. 406.
190 "spins and turns, which": Ibid.
190 "the distance from the earth": Rosen, *Commentariolus*, in Rosen, *TCT*, p. 90.
190 "How exceedingly fine": Copernicus, *ORHS*, p. 27.

14 THE DEBATE OVER THE TWO WORLD SYSTEMS

191 "I know of a modern scientist": Owen Gingerich, "Erasmus Reinhold," *DSB*, 11, p. 366.
192 "Copernicus . . . affirmeth": Kuhn, *The Copernican Revolution*, p. 186.
192 That is trulye to be gathered: Francis R. Johnson, "The Influence of Thomas Digges," p. 396.
192 "more than Herculean": Armitage, *Copernicus and Modern Astronomy*, p. 165.
192 "*Vulgi opinio Error*": Gingerich, *The Book Nobody Read*, p. 119.
193 "so that Englishmen might not": Ibid., p. 119.
193 *Speak, are there many spheres*: Marlowe, *The Tragical History of Dr. Faustus*, act 2, scene 2, ll. 34ff.
193 "There are then innumerable suns": Koyré, *From the Closed World to the Infinite Universe*, p. 49.
194 There are no ends: Ibid., p. 44.
195 "a second Ptolemy": Dreyer, *Tycho Brahe*, p. 74.
196 "I doubted no longer": Ferguson, *Tycho and Kepler*, p. 47.
199 "How many mathematicians are there": A. R. Hall, *The Scientific Revolution*, p. 126.

199 "a still unexhausted treasure": Caspar, *Kepler*, p. 64.
199 The earth's orbit: Owen Gingerich, "Johannes Kepler," *DSB*, 7, p. 290.
200 "Have faith, Galilii": Koestler, *The Sleepwalkers*, p. 364.
201 "a brilliant speculation": Ferguson, *Tycho and Kepler*, p. 255.
201 "You will come not so much as a guest": Owen Gingerich, "Johannes Kepler," *DSB*, 7, p. 293.
202 "I consider it a divine decree": Caspar, *Kepler*, p. 131.
202 "Although he knew": Ferguson, *Tycho and Kepler*, p. 284.
202 "Divine Providence granted us": Owen Gingerich, "Johannes Kepler," *DSB*, 7, p. 295.
203 *Hereafter, when they*: Milton, *Paradise Lost*, book VIII, ll. 79–84.
204 "a book full of astronomical": Owen Gingerich, "Johannes Kepler," *DSB*, 7, p. 297.
204 "perspective glasses": Francis R. Johnson, "The Influence of Thomas Digges," p. 401.
204 a small "telescope": Ibid., p. 403.
205 "Medicean Stars": Galileo, *The Starry Messenger*, in Drake, *Discoveries and Opinions of Galileo*, p. 21.
206 "necessary for the contemplation": Caspar, *Kepler*, p. 296.
206 "That the same thought": Ibid., pp. 276–77.
206 "those opinions of yours": John Donne, "Ignatius His Conclave," in *Complete Poetry and Selected Prose of John Donne*, p. 365.
206 *And new Philosophy*: Donne, "The Anatomy of the World," in *The Complete Poetry of John Donne*, pp. 277–78.
207 *I used to measure the heavens*: Owen Gingerich, "Johannes Kepler," *DSB*, 7, p. 307.
207 "foolish and absurd": Armitage, *Copernicus and Modern Astronomy*, p. 189.
209 "It would still be excessive": Galileo, *Dialogue Concerning the Two Chief World Systems, Ptolemaic and Copernican*, p. 464.
209 "Your Galileo has ventured to meddle": De Santillana, *The Crime of Galileo*, p. 191.
210 "had altogether given rise": Koestler, *The Sleepwalkers*, p. 503.
210 Take note theologians: Galileo, *Dialogue Concerning the Two Chief World Systems, Ptolemaic and Copernican*, p. v.

15 THE SCIENTIFIC REVOLUTION

211 a "hypothesis" . . . "really true": Mary Hesse, "Francis Bacon," *DSB*, 1, p. 372.
212 "*Cogito ergo sum*": Descartes, *PWD*, vol. 1, p. 127.
212 "quantity of motion": Ibid., p. 240.
212 "In the beginning, in his omnipotence": Ibid., p. 240.
212 "Each and every thing": Ibid., p. 240.
212 "all motion is in itself rectilinear": Ibid., pp. 241–42.
212 "If a body collides": Ibid., p. 242.
213 "true mathematics": Michael S. Mahoney, "Descartes: Mathematics and Physics," *DSB*, 4, p. 55.
213 hypothetical "new world": Descartes, *PWD*, vol. 1, p. 90.

214 "almost nothing I can approve as true": Gaukroger, *Descartes: An Intellectual Biography*, p. 421.
214 "Otherwise," he wrote: Desmond M. Clarke, "Pascal's Philosophy of Science," in Hammond, *The Cambridge Companion to Pascal*, p. 114.
216 "All this was in the two plague years": Westfall, *NAR*, p. 143.
217 "rays which make blue": Ibid., p. 160.
217 For the best and safest method: Ibid., p. 242.
218 "extremely well pleased to see": Manuel, *A Portrait of Isaac Newton*, p. 144.
218 "What Descartes did": Westfall, *NAR*, p. 274.
218 "what he thought the Curve would be": Ibid., p. 403.
219 Our present work sets forth: Newton, *MPNP*, p. 382.
219 "quantity of matter": Ibid., p. 403.
219 "quantity of motion": Ibid., p. 404.
220 "inherent force of matter": Ibid.
220 "Impressed force": Ibid., p. 405.
220 "are impelled, or in any way": Ibid., pp. 405–8.
221 Law 1: Every body perseveres: Ibid., pp. 416–17.
222 "by radii drawn to the center": Ibid., p. 797.
222 "Gravity exists in all bodies": Ibid., p. 810.
222 "The planets move in ellipses": Ibid., p. 817.
222 "The axes of the planets": Ibid., p. 821.
222 "the ebb and flow": Ibid., p. 835.
223 "To find the precession": Ibid., p. 885.
223 "the precession of the equinoxes": Ibid., p. 887.
223 "The comets are higher": Ibid., p. 888.
223 "This most elegant system": Ibid., p. 940.
223 "corrected by the author's own hand": Newton, *Opticks*, p. lxxvii.
223 "My design in this Book": Ibid., p. 1.
224 "the oldest and most celebrated": Ibid., p. 369.
224 "All these things being consider'd": Ibid., p. 400.
224 "Light is propagated": Ibid., p. 277.
225 "method of fluxions": Westfall, *NAR*, p. 143.
225 "four metals" of alchemy: A. P. Youschkevitch, "Isaac Newton," *DSB*, 10, p. 81.
225 "He lived honored by his compatriots": Voltaire, *Letters on England*, p. 69.
225 "I do not know what I may appear": Sullivan, *Isaac Newton*, p. 250.

16 SAMARKAND TO ISTANBUL: THE LONG TWILIGHT OF ISLAMIC SCIENCE

227 "absorption of Arabic knowledge": Dampier, *History of Science*, p. 82.
227 "the decline of Arabic": Ibid., p. 82.
228 so-called al-Tusi couple: F. Jamil Ragep, "Copernicus and His Islamic Predecessors," p. 128.
228 Urdi's lemma: Ibid.
232 "passing through the nine": Sayílí, *The Observatory in Islam*, p. 290.
233 "People of learning": Ibid., pp. 292–93.

233 "In the *Zij* of Ulugh Beg": Ibid., p. 293.
233 "The King of Kings": Ibid.
233 "Let it be known": Ihsanoğlu, *STLOE*, vol. 2, pp. 33–34.

17 SCIENCE LOST AND FOUND

238 "There stands a globe": Ovid, *Fasti* VI, 277–80.
239 "Our friend Posidonius as you know": Cicero, *The Nature of the Gods*, II, 87–89.
239 "Archimedes' significance": Clagett, *Archimedes in the Middle Ages*, vol. 1, p. 1.
240 "Unlike the *Elements* of Euclid": Boyer, *A History of Mathematics*, p. 136.
240 Codex A "was the source": Marshall Clagett, "Archimedes," *DSB*, 1, p. 223.
243 "a disastrously misguided attempt": John Lowden, "Archimedes into Icon," p. 236.
245 "I thought fit to write out for you": Heath, *The Works of Archimedes*, Supplement on *The Method of Archimedes*, p. 12.
245 "The cylinder with base equal": Ibid., p. 18.
246 the *Stomachion* "is a kind of game": Dijksterhuis, *Archimedes*, p. 409.
246 "it consisted of fourteen bits of ivory": Ibid., p. 410.
248 Archimedes to Eratosthenes greeting: Heath, *The Works of Archimedes*, Supplement on *The Method of Archimedes*, p. 12.

18 HARRAN: THE ROAD TO BAGHDAD

250 "the city that . . . was ruled": Segal, *Edessa*, p. iii.
251 "the supreme philosopher": Thorndike, *HMES*, vol. 1, p. 661.
252 We are the heirs and offspring of paganism: Rozenfeld and Ihsanoğlu, *MASI*, p. 56.
253 "handbook for manufacturing": Ibid., p. 55.
253 "sorcery and spells": *Thousand Nights and One Night*, vol. 3, p. 382.
253 "the spindle of Necessity": Plato, *Republic* X, 617c.
255 Baghdad, in the heart of Islam: Clot, *Harun al-Rashid*, p. 151.
255 *Behold Baghdad!*: Kennedy, *The Court of the Caliphs*, p. 106.

BIBLIOGRAPHY

DSB: Dictionary of Scientific Biography

Abell, George. *Exploration of the Universe.* 2nd ed. New York, 1969.
Africa, Thomas W. "Copernicus' Relation to Aristarchus and Pythagoras." *Isis* 52, no. 3 (Sept. 1961): 403–9.
Ahmad, S. Makbul. "Al-Idrisi." *DSB*, 7, pp. 7–9.
———. "Al-Masudi." *DSB*, 9, pp. 171–72.
Al-Dabbagh, J. "Banu Musa." *DSB*, 1, pp. 443–46.
Allan, D. J. "Plato." *DSB*, 11, pp. 22–31.
Anawati, George C. "Hunayn ibn Ishaq." *DSB*, 15, pp. 230–48.
———. "Science." *The Cambridge History of Islam.* Vol. 2, pp. 741–80. Cambridge, 1977.
Anbouba, Adel. "Al-Tusi, Sharaf al-Din." *DSB*, 13, pp. 514–17.
Anonymous. *The Thousand Nights and One Night.* 4 vols. Translated by Powys Mathers from the French translation of J. C. Mardrus. London, 1981.
Arberry, J. A., trans. *The Spiritual Physick of Rhazes.* London, 1950.
Aristophanes. Translated by Benjamin Bickley Rogers. 3 vols. London, 1978.
Aristotle. *The Complete Works.* Edited by Jonathan Barnes. 2 vols. Princeton, 1984.
Armitage, Angus. *Copernicus and Modern Astronomy.* New York, 2004.
———. *Sun, stand thou still: the life and works of Copernicus, the astronomer.* New York, 1947.
Armstrong, A. H., ed. *The Cambridge History of Later Greek and Early Medieval Philosophy.* Cambridge, 1967.
Bacon, Francis. *The New Organum and Related Writings.* Edited by Fulton H. Anderson. New York, 1960.
Bailey, Cyril. *The Greek Atomists and Epicurus.* Oxford, 1964.
Baker, Robert H. *Astronomy: An Introduction.* New York, 1930.
Bald, R. C. *John Donne: A Life.* Oxford, 1971.
Barnes, Jonathan. *Early Greek Philosophers.* New York, 1987.
Bevan, Edwyn. *The House of Ptolemy: A History of Egypt Under the Ptolemaic Dynasty.* Chicago, 1968.
Boardman, John. *The Greeks Overseas.* Baltimore, 1964.

Bos, H. J. M. "Christiaan Huygens." *DSB*, 6, pp. 597–612.

Boyer, Carl B. *A History of Mathematics*. 2nd ed. Revised by Uta C. Mertzbach. New York, 1991.

Brockelmann, Carl. *History of the Islamic Peoples*. Translated by Joel Carmichael and Moshe Perlmann. New York, 1960.

Bulmer-Thomas, Ivor. "Euclid." *DSB*, 4, pp. 414–59.

———. "Eutocius of Ascalon." *DSB*, 4, pp. 488–91.

———. "Isidorus of Miletus." *DSB*, 7, pp. 28–30.

———. "Menelaus of Alexandria." *DSB*, 9, pp. 296–302, and 15, pp. 420–21.

———. "Pappus of Alexandria." *DSB*, 10, pp. 293–304.

———. "Theodosius of Bithynia." *DSB*, 13, pp. 319–21.

Burnet, John. *Early Greek Philosophy*. 2nd ed. London, 1908.

———. *Greek Philosophy, Thales to Plato*. London, 1981.

Butterfield, Herbert. *The Origins of Modern Science, 1300–1800*. New York, 1957.

Bylebyl, Jerome J. "William Harvey." *DSB*, 6, pp. 150–62.

Callus, Daniel A., ed. *Robert Grosseteste, Scholar and Bishop*. Oxford, 1955.

Carmody, Francis J. *The Astronomical Works of Thabit Ibn Qurra*. Berkeley, 1960.

Casper, Max. *Kepler*. Translated by C. Doris Hellman. New York, 1962.

Casson, Lionel. *Libraries in the Ancient World*. New Haven, 2001.

Chatillon, Jean. "Giles of Rome." *DSB*, 5, pp. 402–3.

Cicero. *De Republica*. Translated by Clinton Walker Keyes. London, 1928.

———. *The Nature of the Gods*. Translated by H. C. P. McGregor. Harmondsworth, 1972.

Clagett, Marshall. "Adelard of Bath." *DSB*, 1, pp. 61–64.

———. "Archimedes." *DSB*, 1, pp. 213–31.

———, ed. *Archimedes in the Middle Ages*. Vol. 1, *The Arabo-Latin Tradition*. Madison, Wisc., 1964.

———. *Greek Science in Antiquity*. London, 2001.

———. "John of Palermo." *DSB*, 7, pp. 133–34.

———. "Nicole Oresme." *DSB*, 10, pp. 223–30.

———. *The Science of Mechanics in the Middle Ages*. Madison, Wisc., 1959.

Clot, André. *Harun al-Rashid and the World of the Thousand and One Nights*. Translated by John Howe. London, 2005.

Cobban, Alan B. *The Medieval Universities: Their Development and Organization*. London, 1975.

Cohen, I. Bernard. *Revolution in Science*. Cambridge, Mass., 1985.

Cohen, Morris, and I. E. Drabkin. *Source Book in Greek Science*. Cambridge, Mass., 1958.

Collins, Roger. *Early Medieval Europe, 300–1100*. Basingstoke, 1991.

———. *Early Medieval Spain: Unity in Diversity, 400–1000*. New York, 1995.

Cook, John M. *The Greeks in Ionia and the East*. New York, 1963.

Copernicus, Nicholas. *On the Revolutions of Heavenly Spheres*. Translated by Charles Glen Wallis. Edited by Stephen Hawking. Philadelphia, 2002.

Cottingham, John. *Descartes*. Oxford, 1986.

———, ed. *The Cambridge Companion to Descartes*. Cambridge, 1992.

Cottingham, John, Robert Stoothoff, and Dugald Murdoch. *The Philosophical Writings of Descartes*. 2 vols. Cambridge, 1985.

Crombie, A. C. *Medieval and Early Modern Science*. 2 vols. 2nd ed. Cambridge, Mass., 1973.

———. *Robert Grosseteste and the Origins of Experimental Science, 1100–1700.* Oxford, 1953.

———, ed. *Scientific Change: Historical Sketches in the Intellectual, Social and Technical Conditions for Scientific Discovery and Technical Invention from Antiquity to the Present.* New York, 1963.

Crombie, A. C., Michael S. Mahone, and Theodore M. Brown. "René du Perron Descartes." *DSB*, 4, pp. 51–65.

Crombie, A. C., and J. D. North. "Roger Bacon." *DSB*, 1, pp. 377–85.

Dales, Richard. *The Scientific Achievements of the Middle Ages.* Philadelphia, 1973.

Daly, John F. "Johannes de Sacrobosco (John of Holywood). *DSB*, 12, pp. 60–63.

Dampier, William Cecil. *A History of Science and Its Relation with Philosophy and Religion,* 3rd ed. New York, 1944.

Dannenfeldt, Karl H. "Hermes Trismegistus." *DSB*, 6, pp. 305–6.

Dante. *The Divine Comedy.* Translated by H. F. Cary. New York, 1908.

Davidson, Herbert A. *Alfarabi, Avicenna, and Averroes, on Intellect.* New York, 1992.

De Santillana, Giorgio. *The Crime of Galileo.* Chicago, 1955.

———. *The Origins of Scientific Thought: Anaximander to Proclus, 600 B.C. to A.D. 300.* Chicago, 1961.

Descartes, René. *The Philosophical Writings of Descartes.* 3 vols. Translated by John Cottingham, Robert Stoothoff, and Dugald Murdoch. Cambridge, 1985–91.

Dicks, D. R. *Early Greek Astronomy to Aristotle.* Ithaca, N.Y., 1970.

———. "Eratosthenes." *DSB*, 4, pp. 388–93.

———. "Hecataeus." *DSB*, 6, pp. 212–13.

Dictionary of Scientific Biography (DSB). 16 vols. Edited by Charles Coulston Gillespie. New York, 1970–80.

Dijksterhuis, F. J. *Archimedes.* Princeton, 1987.

Dilgan, Hamit. "Qadi Zada al-Rumi." *DSB*, 11, pp. 227–29.

Diogenes Laertius. *Lives of Eminent Philosophers.* 2 vols. Translated by H. D. Hicks. Cambridge, Mass., 1925.

Dizer, Muammer, ed. *International Symposium on the Observatory in Islam, 19–23 September 1977.* Istanbul, 1980.

Dold-Samplonius, Yvonne, and Heinrich Hermelink. "Al-Jayyani." *DSB*, 7, pp. 82–83.

Donne, John. *The Complete Poetry of John Donne.* Edited by John T. Shawcross. London, 1968.

———. *The Complete Poetry and Selected Prose of John Donne.* Edited by John Hayward. London, 1929.

Dozy, Reinhart. *A History of the Moslems in Spain.* Translated by Frances Griffin Stokes. London, 1913.

Drachmann, A. G. "Ctesibius." *DSB*, 3, pp. 491–92.

———. "Hero of Alexandria." *DSB*, 6, pp. 310–15.

———. "Philo of Byzantium." *DSB*, 10, pp. 586–89.

Drake, Stillman, trans. *Discoveries and Opinions of Galileo.* Garden City, N.Y., 1952.

———. "Galileo Galilei," *DSB*, 5, pp. 237–48.

Dreyer, J. L. E. *A History of Astronomy from Thales to Kepler.* New York, 1953.

———. *Tycho Brahe.* New York, 1963.

Dunlop, D. M. *Arab Science in the West.* Karachi, 1958.

Easton, Joy B. "John Dee." *DSB*, 4, pp. 5–6.

———. "Thomas Digges." *DSB*, 4, pp. 97–98.

————. "Robert Recorde." *DSB*, 11, pp. 338–40.

Edmonds, J. M., trans. *Lyra Graeca*. 3 vols. London, 1928.

Eicholz, David E. "Pliny." *DSB*, 11, pp. 38–40.

Encyclopedia of the History of Arabic Science. 3 vols. Edited by Roshdi Rashed and Regis Morelon. London, 1996.

Encyclopedia of the Scientific Revolution. Edited by William Applebaum. New York, 2000.

Esposito, John L., ed. *The Oxford History of Islam*. Oxford, 1999.

Evans, James. *The History and Practice of Ancient Astronomy*. Oxford, 1998.

Fakhry, Majid. *A History of Islamic Philosophy*. 3rd ed. Chicago, 1992.

Farrington, Benjamin. *Greek Science: Its Meaning for Us*. Baltimore, 1953.

Ferguson, Kitty. *Tycho and Kepler: The Unlikely Partnership That Forever Changed Our Understanding of the Heavens*. New York, 2002.

Fisher, W. B., ed. *The Cambridge History of Iran*. vol. 7. Cambridge, 1968–81.

Fletcher, Richard. *Moorish Spain*. London, 1992.

Freely, John. *The Emergence of Modern Science, East and West*. Istanbul, 2004.

Freeman, Kathleen. *The Presocratic Philosophers*. Oxford, 1953.

Fritz, Kurt von. "Philolaus of Crotona." *DSB*, 10, pp. 589–91.

————. "Pythagoras of Samos." *DSB*, 11, pp. 219–25.

————. "Zeno of Elea." *DSB*, 14, pp. 607–12.

Fryde, Edmund. *The Early Palaeologan Renaissance, 1261–ca. 1360*. Leiden, 2000.

Furley, David J. "Epicurus." *DSB*, 4, pp. 381–82.

————. "Heraclitus of Ephesus." *DSB*, 6, pp. 289–91.

————. "Lucretius." *DSB*, 8, pp. 536–39.

Gade, John A. *The Life and Times of Tycho Brahe*. Princeton, 1947.

Galilei, Galileo. *Dialogue Concerning the Two Chief World Systems, Ptolemaic and Copernican*. Translated by Stillman Drake. Berkeley, 1967.

————. *Discourses Concerning Two New Sciences, of Mechanics and of Motion*. New York, 1914.

Gassendi, Pierre. *The Life of Copernicus (1473–1543)*. With notes by Olivier Thill. Fairfax, Va., 2003.

Gaukroger, Stephen. *Descartes: An Intellectual Biography*. Oxford, 1995.

Geanakoplos, Dino John. *Greek Scholars in Venice: Studies in the Dissemination of Greek Learning from Byzantium to Western Europe*. Cambridge, Mass., 1962.

Geymont, Ludovico. *Galileo: A Biography and Inquiry into His Philosophy of Science*. New York, 1965.

Gingerich, Owen. *The Book Nobody Read: Chasing the Revolutions of Nicolaus Copernicus*. New York, 2004.

————. "Johannes Kepler." *DSB*, 7, pp. 289–312.

————. "Erasmus Reinhold." *DSB*, 11, pp. 365–67.

Gliozzi, Marion. "Evangelista Torricelli." *DSB*, 13, pp. 433–40.

Goldstein, Bernard R. "The Heritage of Arabic Science in Hebrew." In *Encyclopedia of the History of Arabic Science*, vol. 1, pp. 276–83.

Goldstein, Bernard R., and Alan C. Bowen. "A New View of Early Greek Astronomy." *Isis* 74, no. 3 (Sept. 1983): 330–40.

Gottschalk, H. B. "Strato of Lampsacus." *DSB*, 13, pp. 91–95.

Grant, Edward. "Jordanus de Nemore." *DSB*, 7, 171–79.

————. "Peter Peregrinus." *DSB*, 10, 532–40.

————. *Physical Science in the Middle Ages.* New York, 1971.

Gregory, John. *The Neoplatonists: A Reader.* London, 1999.

Griffiths, Robert I. "Was There a Crisis Before the Copernican Revolution? A Reappraisal of Gingerich's Criticisms of Kuhn." In *Proceedings of the Biennial Meeting of the Philosophy of Science Association.* Vol. 1., pp. 127–36. Chicago, 1988.

Guerlac, Henry. "Copernicus and Aristotle's Cosmos." *Journal of the History of Ideas* 29, no. 1 (1968): 109–13.

Gutas, Dimitri. *Greek Philosophers in the Arabic Tradition.* Aldershot, U.K., 2000.

————. *Greek Thought, Arabic Culture: The Graeco-Arabic Translation Movement in Baghdad and Early Abbasid Society.* London, 1998.

Guthrie, William K. C. *A History of Greek Philosophy.* 6 vols. Cambridge, 1962–81.

Hall, A. Rupert. *The Scientific Revolution, 1500–1800.* Boston, 1956.

Hall, Marie Boas. *Robert Boyle and Seventeenth-Century Chemistry.* Cambridge, 1958.

Hall, Robert E. "Al-Khazini." *DSB,* 7, pp. 335–51.

Hamarneh, Sami. "Al-Majusi." *DSB,* 9, pp. 40–42.

————. "Al-Zahrawi." *DSB,* 14, pp. 584–85.

————. "Ibn Zuhr." *DSB,* 14, pp. 637–39.

Hammond, Nicholas, ed. *The Cambridge Companion to Pascal.* Cambridge, 2003.

Hammond, N. G. I., and H. H. Scullard, eds. *The Oxford Classical Dictionary.* 2nd ed. Oxford, 1970.

Hartner, Willy. "Al-Battani." *DSB,* 1, pp. 507–16.

Harvey, E. Ruth. "Qusta ibn Luqa." *DSB,* 11, pp. 244–46.

Haskins, Charles Homer. *The Renaissance of the Twelfth Century.* New York, 1957.

————. *The Rise of Universities.* New York, 1923.

————. *Studies in the History of Mediaeval Science.* Cambridge, Mass., 1924.

Heath, Thomas L. *Aristarchus of Samos, the Ancient Copernicus.* Oxford, 1959.

————. *Diophantus of Alexandria: A Study in the History of Greek Algebra.* New York, 1964.

————. *A History of Greek Mathematics.* 2 vols. Oxford, 1921.

————. *A Manual of Greek Mathematics.* New York, 1963.

————. *The Thirteen Books of Euclid's Elements.* New York, 1956.

————. *The Works of Archimedes.* New York, 1912.

Heiberg, Johann Ludwig. *Mathematics and Physical Science in Classical Antiquity.* Translated by D. C. Macgregor. London, 1922.

Heidel, W. A. *The Heroic Age of Science.* Baltimore, 1933.

Hellman, C. Doris. "Tycho Brahe." *DSB,* 2, pp. 401–14.

Hellman, C. Doris, and Noel M. Swerdlow. "Georg Peurbach." *DSB,* 15, pp. 473–79.

Henry, John. *The Scientific Revolution and the Origins of Modern Science.* New York, 2002.

Herodotus. *The Histories.* Translated by Aubrey de Sélincourt. Harmondsworth, 1954.

Hesiod, the Homeric Hymns and Homerica. Translated by Hugh G. Evelyn-White. London, 1926.

Hesse, Mary. "Francis Bacon." *DSB,* 1, pp. 372–77.

Hill, Donald R. "Al-Jazari." *DSB,* 15, pp. 253–55.

————. *Islamic Sciences and Engineering.* Edinburgh, 1993.

Hirst, Anthony, and Michael Silk. *Alexandria, Real and Imagined.* Aldershot, U.K., 2004.

Hofmann, Joseph E. "Nicholas Cusa." *DSB,* 3, pp. 512–16.

———. "Gottfried Wilhelm Leibniz." *DSB*, 8, pp. 149–68.

Holt, R. M., Ann K. S. Lambton, and Bernard Lewis, eds. *The Cambridge History of Islam*. Cambridge, 1977.

Hourani, George F. *Averroes on the Harmony of Religion and Philosophy*. London, 1961.

———. "Ibn Tufayl." *DSB*, 13, pp. 488–89.

Huff, Toby E. *The Rise of Early Modern Science: Islam, China and the West*. Cambridge, 1993.

Hugonnard-Roche, Henri. "The Influence of Arabic Astronomy in the Medieval West." In *Encyclopedia of the History of Arabic Science*, vol. 1, pp. 284–85.

Hussey, Edward. *The Presocratics*. New York, 1972.

Huxley, G. L. "Anthemius of Tralles." *DSB*, 1, pp. 169–70.

———. *The Early Ionians*. London, 1966.

———. "Eudoxus of Cnidus." *DSB*, 4, pp. 465–67.

Ierodiakonou, Katerina, ed. *Byzantine Philosophy and Its Ancient Sources*. Oxford, 2002.

Ihsanoğlu, Ekmeleddin. *History of the Ottoman State, Society and Civilization*. Istanbul, 2001.

———. *Science, Technology and Learning in the Ottoman Empire*. Istanbul, 1992.

———, ed. *Transfer of Modern Science and Technology to the Muslim World: Proceedings of the International Symposium on Modern Sciences and the Modern World*. Istanbul, 2–4 September 1987. Istanbul, 1992.

Irbie-Massie, Georgia L., and Paul T. Keyser. *Greek Science of the Hellenistic Era: A Sourcebook*. London, 2002.

Iskandar, Albert Z. "Hunayn Ibn Ishaq." *DSB*, 15, pp. 230–49.

———. "Ibn al-Nafis." *DSB*, 9, pp. 602–6.

———. "Ibn Rushd (Averroës)." *DSB*, 12, pp. 12, 1–9.

———. "Ibn Sina (Avicenna)." *DSB*, 15, pp. 494–501.

Jayussi, Salma Khadra, ed. *The Legacy of Muslim Spain*. 2 vols. Leiden, 1992.

Jerusalem Bible. Popular edition. London, 1974.

Johnson, Francis R. *Astronomical Thought in Renaissance England*. Baltimore, 1937.

———. "The Influence of Thomas Digges in the Progress of Modern Astronomy in Sixteenth-Century England." *Osiris* 1 (June 1936): 390–410.

———. "Marlowe's Astronomy and Renaissance Skepticism." *Journal of English Literary History* 3, no. 4 (Dec. 1946): 241–54.

Johnson, Paul. *A History of the Jews*. London, 1995.

Jolivet, Jean, and Roshdi Rashed. "Al-Kindi." *DSB*, 15, pp. 261–67.

Joly, Robert. "Hippocrates of Cos." *DSB*, 6, pp. 418–31.

Jones, Charles W. "The Venerable Bede" *DSB*, 1, pp. 564–66.

Kaler, James B. *Astronomy!* New York, 1994.

Kantorowicz, Ernst. *Frederick the Second, 1194–1250*. Translated by E. O. Lorimer. New York, 1931.

Kari-Niazov, T. N. "Ulugh Beg." *DSB*, 13, pp. 535–37.

Kelly, Suzanne. "William Gilbert." *DSB*, 5, pp. 396–401.

Kennedy, E. S. "Al-Battani." *DSB*, 1, pp. 507–16.

———. "Al-Biruni." *DSB*, 2, pp. 147–58.

———. "The Exact Sciences." In *The Cambridge History of Iran*, 4: Cambridge, 1968, pp. 378–95.

Kennedy, E. S., et al. *Studies in the Islamic Exact Sciences*. Beirut, 1983.

Kennedy, Hugh. *The Court of the Caliphs: The Rise and Fall of Islam's Greatest Dynasty*. London, 2004.

Kerferd, G. B. "Democritus." *DSB*, 4, pp. 30–35.

Khayyam, Omar. *The Rubaiyat*. 5th ed. Translated by Edward FitzGerald. New York, 1966.

King, David A. "Ibn al-Shatir." *DSB*, 12, pp. 357–64.

———. "Ibn Yunus." *DSB*, 14, pp. 574–80.

Kirk, G. S., and J. E. Raven. *The Presocratic Philosophers*. Cambridge, 1962.

Kline, Morris. *Mathematical Thought from Ancient to Modern Times*. 3 vols. New York, 1990.

Knobel, E. B., ed. *Ulugh Bey's Catalogue of Stars*. Washington, D.C., 1917.

Koestler, Arthur. *The Sleepwalkers: A History of Man's Changing Vision of the Universe*. London, 1959.

Kopal, Zdenek. "Olaus Roemer." *DSB*, 11, pp. 525–27.

The Koran: The Meaning of the Glorious Koran, trans. Mohammed Marmaduke Pickthall. New York, 1953.

Koyré, Alexandre. *The Astronomical Revolution: Copernicus, Kepler, Borelli*. Translated by R. E. W. Madison. Ithaca, N.Y., 1973.

———. *From the Closed World to the Infinite Universe*. New York, 1958.

———. *Newtonian Studies*. Cambridge, Mass., 1965.

Krafft, Fritz. "Otto von Guericke." *DSB*, 5, pp. 574–76.

Kramer, Edna E. "Hypatia." *DSB*, 6, pp. 615–16.

Kren, Claudia. "Dominicus Gundissalinus." *DSB*, 5, pp. 591–93.

———. "Hermann the Lame." *DSB*, 6, pp. 301–3.

———. "Roger of Hereford." *DSB*, 11, pp. 503–4.

Kudlien, Fridolf, and Leonard G. Wilson. "Galen." *DSB*, 5, pp. 227–37.

Kuhn, Thomas S. *The Copernican Revolution: Planetary Astronomy in the Development of Western Thought*. Cambridge, Mass., 1957.

———. *The Structure of Scientific Revolutions*. Chicago, 1976.

Kunitzsch, Paul. "Al-Sufi." *DSB*, 13, pp. 149–50.

Lapidus, Ira. *A History of Islamic Societies*. 2nd ed. Cambridge, 2002.

Lemay, Richard. "Gerard of Cremona." *DSB*, 15, pp. 173–92.

Levey, Martin. "Abraham Bar Hiyya Ha-Nasi (Savasorda)." *DSB*, 1, pp. 22–23.

Lindberg, David C. *The Beginnings of European Science: the European Scientific Tradition in Philosophical, Religious and Institutional Context, 600 B.C. to A.D. 1450*. Chicago, 1992.

———. *Studies in the History of Medieval Optics*. London, 1983.

———. *Theories of Vision from al-Kindi to Kepler*. Chicago, 1976.

———. "Witelo." *DSB*, 14, pp. 457–62.

Lindberg, David C., and Robert S. Westman, eds. *Reappraisal of the Scientific Revolution*. Cambridge, 1990.

Little, A. G., ed. *Roger Bacon: Essays*. Oxford, 1914.

Lloyd, G. E. R. *Early Greek Science: Thales to Aristotle*. New York, 1970.

———. *Greek Science After Aristotle*. New York, 1973.

Lohne, J. A. "Thomas Harriot." *DSB*, 6, pp. 124–29.

Long, A. A., ed. *The Cambridge Companion to Early Greek Philosophy*. Cambridge, 1999.

Longrigg, James. "Anaxagoras." *DSB*, 1, pp. 149–50.

———. "Thales." *DSB*, 13, pp. 295–98.

Lorch, R. P. "Jabir ibn Aflah." *DSB*, 7, pp. 37–39.

Lorris, Guillaume de, and Jean de Meun. *The Romance of the Rose*. Translated by Harry W. Robbins. New York, 1962.

Lowden, John. "Archimedes into Icon: Forging an Image of Byzantium." In *Icons and Word: The Power of Images in Byzantium*. Edited by Antony Eastmond and Liz James. Burlington, Vt., 2003.

Lucretius. *De Rerum Natura (On the Nature of the Universe)*. Translated by R. E. Latham. Revised by John Godwin. London, 1994.

Machamer, Peter, ed. *The Cambridge Companion to Galileo*. Cambridge, 1996.

MacLeod, Roy, ed. *The Library of Alexandria, Centre of Learning in the Ancient World*. London, 2000.

Mahdi, Muhsin. "Al-Farabi." *DSB*, 4, pp. 523–26.

Mahoney, Michael S. "Hero of Alexandria." *DSB*, 6, pp. 310–15.

———. "Pierre de Fermat." *DSB*, 4, 566–76.

Makdisi, George. *The Rise of Colleges: Institutions of Learning in Islam and the West*. Edinburgh, 1981.

———. *The Rise of Humanism in Classical Islam and the West*. Edinburgh, 1990.

Manuel, Frank E. *A Portrait of Isaac Newton*. Cambridge, Mass., 1968.

Marlowe, Christopher. *The Complete Plays*. Edited by Frank Romany and Robert Lindsey. London, 2003.

Masson, Georgina. *Frederich II of Hohenstaufen: A Life*. London, 1957.

Matthew, Donald. *The Norman Kingdom of Italy*. Cambridge, 1992.

McDiarmid, J. B. "Theophrastus." *DSB*, 13, pp. 328–34.

McKeon, Richard. *The Basic Works of Aristotle*. New York, 1941.

McVaugh, Michael. "Constantine the African." *DSB*, 3, pp. 393–95.

Menocal, Maria Rosa. *The Ornament of the World: How Muslims, Jews, and Christians Created a Culture of Tolerance in Medieval Spain*. Boston, 2002.

Menocal, Maria Rosa, et al., eds. *The Literature of Al-Andalus*. New York, 2000.

Merlan, Philip. "Alexander of Aphrodisias." *DSB*, 1, pp. 117–20.

———. "Ammonius." *DSB*, 1, p. 137.

Milton, John. *Paradise Lost*. Edited by Edward Le Comte. New York, 1961.

———. *Paradise Regained*. Edited by Merritt Y. Hughes. New York, ca. 1937.

Minio-Paluello, Lorenzo. "Boethius." *DSB*, 2, pp. 228–36.

———. "James of Venice." *DSB*, 7, pp. 65–67.

———. "Plato of Tivoli." *DSB*, 11, pp. 31–33.

———. "Michael Scott." *DSB*, 9, pp. 361–65.

———. "William of Moerbeke." *DSB*, 9, pp. 434–40.

Monfasani, John. *Byzantine Scholars in Renaissance Italy: Cardinal Bessarion and Other Émigrés; Selected Essays*. Brookfield, Vt., 1995.

———. *Greeks and Latins in Renaissance Italy: Studies in Humanism and Philosophy in the 15th Century*. Brookfield, Vt., ca. 2004.

Montgomery, Scott L. *Science in Translation: Movements of Knowledge Through Cultures and Time*. Chicago, 2000.

Moody, Ernest A. "Jean Buridan." *DSB*, 2, pp. 603–8.

Morelon, Regis. "Eastern Arabic Astronomy Between the Eighth and the Eleventh Centuries." In *Encyclopedia of the History of Arabic Science*, vol. 1, pp. 20–57.

———. "General Survey of Arabic Astronomy." *Encyclopedia of the History of Arabic Science*, vol. 1, pp. 1–19.

Morford, Mark. *The Roman Philosophers: From the Time of Cato the Censor to the Death of Marcus Aurelius*. London, 2002.

Morrow, Glenn R. "Proclus." *DSB*, 11, pp. 160–62.

Mourelatos, Alexander P. D. "Empedocles of Acragas." *DSB*, 4, pp. 367–69.

Murdoch, John. "Euclid." *DSB*, 4, pp. 414–59.

———. "Thomas Bradwardine." *DSB*, 2, pp. 390–97.

Murdoch, John, and Edith Dudley Sylla. "Richard Swineshead." *DSB*, 13, pp. 184–213.

Nasr, Seyyid Hossein. *Islamic Science: An Illustrated History*. Cairo, 1976.

———. "Qutb al-Din Al-Shirzai." *DSB*, 11, pp. 247–53.

———. *Science and Civilization in Islam*. Cambridge, Mass., 1968.

———. "Nasir al-Din al-Tusi." *DSB*, 13, pp. 508–14.

Nasr, Seyyid Hossein, and Oliver Leaman, eds. *History of Islamic Philosophy*. London, 1996.

Netz, Reviel, Fabio Acerbi, and Nigel Wilson. "Towards a Reconstruction of Archimedes' Stomachion." *Sciamus* 5 (2004): 67–99.

Neugebauer, Otto. *The Astronomical Tables of Al-Khwarizmi*. Copenhagen, 1962.

———. *The Exact Sciences in Antiquity*. 2nd ed. Providence, R.I., 1957.

———. *A History of Ancient Mathematical Astronomy*. 3 vols. New York, 1975.

Newton, Isaac. *Mathematical Principles of Natural Philosophy*. Translated by I. Bernard Cohen and Anne Whitman. Berkeley, 1999.

———. *Opticks, or A Treatise of the Reflections, Refractions, Inflections, and Colours of Light*. London, 1952. (Reprint of the 4th ed., London, 1730.)

North, John David. *The Norton History of Astronomy and Cosmology*. New York, 1995.

———. "Richard of Wallingford." *DSB*, 11, pp. 414–16.

O'Leary, De Lacy. *Arabic Thought and Its Place in History*. London, 1939.

———. *How Greek Science Passed to the Arabs*. London, 1949.

Omar, Saleh Beshara. *Ibn al-Haytham's Optics: A Study of Experimental Science*. Minneapolis, 1977.

O'Neil, W. M. *Early Astronomy, from Babylonia to Copernicus*. Sydney, 1986.

Osborne, Robin. *Greece in the Making, 1200–479 B.C.* London, 1996.

Osler, Margaret, ed. *Rethinking the Scientific Revolution*. New York, 2000.

Ovid. *Fasti*. Translated by James George Frazer. London, 1959.

Owen, G. E. L., D. M. Balme, Leonard G. Wilson, and L. Minio-Paluello. "Aristotle." *DSB*, 1, pp. 250–81.

Pannekoek, A. *A History of Astronomy*. New York, 1961.

Pastor, Ludwig. *The History of the Popes*. 5th ed. Vol. 6. Edited by F. I. Antrobus. London, 1950.

Pausanias. *Description of Greece*. 2 vols. Translated by Peter Levi. Harmondsworth, 1985.

Payne-Gaposchkin, Cecelia. *Introduction to Astronomy*. New York, 1954.

Permuda, Loris. "Pietro d'Abano." *DSB*, 1, pp. 4–5.

Peters, C. H. F., and E. B. Knobel. *Ptolemy's Catalogue of the Stars: A Revision of the Almagest*. Washington, D.C., 1915.

Peters, Francis E. *Allah's Commonwealth: A History of Islam in the Near East, 600–1100*. New York, 1973.

———. *Aristotle and the Arabs: the Aristotelian Tradition in Islam*. New York, 1968.

———. *The Harvest of Hellenism: A History of the Near East from Alexander the Great to the Triumph of Christianity*. New York, 1970.

Petry, Carl F., ed. *The Cambridge History of Egypt*. Vol. 1, *Islamic Egypt, 640–1517*. Cambridge, 1998.

Pines, Shlomo. "Ibn Bajja." *DSB*, 1, pp. 408–10.

———. "Moses Maimonides." *DSB*, 9, pp. 27–32.

———. "Al-Razi (Rhazes)." *DSB*, 11, pp. 323–26.

———. "What Was Original in Arabic Science." In *Scientific Change*, edited by A. C. Crombie, pp. 206–18.

Pingree, David. "Brahmagupta." *DSB*, 2, pp. 416–18.

———. "Al-Fazari." *DSB*, 4, pp. 555–56.

———. "The Greek Influence on Early Islamic Mathematical Astronomy." *Journal of the American Oriental Society* 93 no. 1 (1973): 32–43.

———. "Gregory Chioniades and Palaeologan Astronomy." *Dumbarton Oaks Papers* 18 (1964): 133–60.

———. "Leo the Mathematician." *DSB*, 8, pp. 190–92.

———. "Masha'allah." *DSB*, 9, pp. 159–62.

Plato. *Complete Works*. Edited by John M. Cooper. Indianapolis, 1997.

Pliny the Elder. *Natural History*. 10 vols. Translated by H. Rackham et al. Cambridge, Mass., 1942–63.

Plutarch. *Plutarch's Lives*. 10 vols. Translated by Bernadotte Perrin. Cambridge, Mass., 1914–26.

———. *Plutarch's Moralia*. 15 vols. Translated by Harold Cherniss and William C. Helmbold. Cambridge, Mass., 1957.

Poulle, Emmanuel. "John of Sicily." *DSB*, 7, pp. 141–42.

———. "William of St. Cloud." *DSB*, 14, pp. 389–91.

Price, Derek De Solla. "An Ancient Greek Computer." *Scientific American* (1959): 60–67.

Ptolemy's Almagest. Translated and annotated by G. J. Toomer, with a foreword by Owen Gingerich. Princeton, 1998.

Quandt, Abigail. "The Archimedes Palimpsest: Conservation Treatment, Digital Imaging and Transcription of a Rare Medieval Manuscript." In *Works of Art on Paper: Books, Documents and Photographs; Contributions to the Baltimore Congress, 2–6 September 2002*. Edited by Vincent Daniels, Alan Donnithorne, and Perry Smith, pp. 165–70. London, 2002.

Ragep, F. Jamil. " 'Ali Qushji and Regiomontanus: Eccentric Transformations and Copernican Revolutions." *Journal for the History of Astronomy* 36 (2005): pp. 359–71.

———. "Copernicus and His Islamic Predecessors: Some Historical Remarks." *Filozofski Vestnik* 25, no. 2 (2004): pp. 125–42.

Ragep, F. Jamil, and Sally P. Ragep, eds. *Tradition, Transmission, Transformation*. Leiden, 1996.

Rashdall, Hastings. *The Universities in Europe in the Middle Ages*. 3 vols. London, 1936.

———. "Kamal al-Din al-Farisi." *DSB*, 7, pp. 212–19.

Rashed, Roshdi, and B. Vahadzadeh. *Omar Khayyam, the Mathematician*. Winona Lake, Ind., ca. 2000.

Read, Jan. *The Moors in Spain and Portugal*. London, 1974.

Riddle, John M. "Dioscorides." *DSB*, 4, pp. 119–23.

Rihill, T. E. *Greek Science*. Oxford, 1999.

Rochot, Bernard. "Pierre Gassendi," *DSB*, 5, 284–90.

Ronan, Colin A. *The Cambridge Illustrated History of the World's Science*. London, 1983.

———. "Edmond Halley." *DSB*, 6, pp. 67–74.

Rosen, Edward. "Federico Commandino." *DSB*, 3, pp. 363–65.

———. "Nicholas Copernicus." *DSB*, 3, pp. 401–11.
———. "Andreas Osiander." *DSB*, 10, pp. 245–46.
———. "Johannes Regiomontanus." *DSB*, 11, pp. 348–52.
———. "George Joachim Rheticus." *DSB*, 11, pp. 395–98.
———. *Three Copernican Treatises.* New York, 1959.
———. "Was Copernicus a Neoplatonist?" *Journal of the History of Ideas* 44, no. 4 (Oct. 1983): 667–69.
Rosenfeld, B. A., and A. T. Grigorian. "Thabit ibn Qurra." *DSB*, 8, pp. 288–95.
Rosenthal, Franz. *The Classical Heritage in Islam.* Translated by Emile and Jenny Marmorstein. London, 1975.
———. *Greek Philosophy and the Arabs.* Brookfield, Vt., 1990.
———. *Science and Medicine in Islam.* Brookfield, Vt., 1990.
Rosinska, Grazyna. "Nasir al-Din al-Tusi and Ibn al-Shatir in Cracow?" *Isis* 65, no. 2 (1974): pp. 239–43.
Ross, David. *Aristotle.* 6th ed. London, 1995.
Rozenfeld, B. A., and Ekmeleddin Ihsanoğlu. *Mathematicians, Astronomers and Other Scholars of Islamic Civilisations and Their Works (7th–14th Centuries).* Istanbul, 2003.
Rudnicki, Jozef. *Nicolaus Copernicus.* London, 1943.
Runciman, Steven. *The Last Byzantine Renaissance.* Cambridge, 1970.
Sabra, A. I. "The Appropriation and Subsequent Naturalization of Greek Science in Medieval Islam: A Preliminary Statement." In *Tradition, Transmission, Transformation.* Edited by F. Jamil Ragep and Sally B. Ragep, pp. 3–27.
———. "Al-Farghani." *DSB*, 4, pp. 541–45.
———. "Ibn al-Haytham." *DSB*, 6, pp. 189–210.
———. "Al-Jawhari." *DSB*, 7, pp. 79–80.
———. *Theories of Light, from Descartes to Newton.* London, 1967.
Sa'di of Shiraz. *Tales from the Bustun . . . and Gulistan.* Translated by Reuben Levy. London, 1928.
Said, Edward W. *Culture and Imperialism.* New York, 1993.
Saidan, A. S. "Al-Qalasadi." *DSB*, 11, pp. 229–30.
Saliba, George. *A History of Arabic Astronomy.* New York, 1994.
Sambursky, Samuel. "John Philoponus." *DSB*, 7, pp. 134–39.
———. *The Physical World of the Greeks.* Translated by Merton Dagut. London, 1956.
———. *Physics of the Stoics.* New York, 1959.
Samso, Julio. "Al-Bitruji." *DSB*, 15, pp. 33–36.
———. "Levi ben Gerson." *DSB*, 8, pp. 279–82.
Sandys, J. F., *A History of Classical Scholarship.* Reprint, New York, 1964.
Sarton, George. *Galen of Pergamum.* Lawrence, Kans., 1954.
———. *A History of Science.* 2 vols. Cambridge, Mass., 1952.
———. *Introduction to the History of Science.* 3 vols. Baltimore, 1927–48.
Sayílí, Aydín. *The Observatory in Islam and Its General Place in the History of the Observatory.* Ankara, 1960.
Sears, Francis Weston. *Optics.* 3rd ed. Reading, Mass., 1958.
Segal, J. B. *Edessa, "The Blessed City."* Oxford, 1970.
Seneca. *Natural Questions.* Translated by J. Clarke. London, 1910.
Sesiano, Jacques. "Diophantus of Alexandria." *DSB*, 15, 118–22.
Ševčenko, I. "Theodore Metochites, the Chora and the Intellectual Trends of His Time." In *Kariye Djami.* Edited by Paul Underwood, vol. 4, pp. 19–91.

Sharpe, William D. "Isidore of Seville." *DSB*, 7, pp. 28–30.

Solmsen, F. *Aristotle's System of the Physical World: A Comparison with His Predecessors.* Ithaca, N.Y., 1960.

Spenser, Edmund. *The Faerie Queene.* 2 vols. New York, 1927.

Stahl, William H. "Aristarchus of Samos." *DSB*, 1, pp. 246–50.

———. "The Greek Heliocentric Theory and Its Abandonment." *Transactions and Proceedings of the American Philological Society* 76 (1945): 321–32.

———. "Macrobius." *DSB*, 9, pp. 1–2.

———. "Martianus Capella." *DSB*, 9, pp. 140–41.

Stern, S. M. "Isaac Israeli." *DSB*, 7, pp. 22–23.

Strabo. *Geography.* Translated by Horace Leonard Jones. 8 vols. Cambridge, Mass., 1982.

Struik, D. J. "Gerbert d'Aurillac." *DSB*, 5, pp. 364–66.

Sullivan, J. W. N. *Isaac Newton, 1642–1727.* New York, 1938.

Swerdlow, N. M., and O. Neugebauer. *Mathematical Astronomy in Copernicus' De Revolutionibus,* 2 vols. New York, 1984.

Talbot, Charles H. "Stephen of Antioch." *DSB*, 13, pp. 38–39.

Taran, Leonardo. "Anaximander." *DSB*, 1, pp. 150–51.

———. "Anaximenes." *DSB*, pp. 151–52.

———. "Aratus of Soli." *DSB*, pp. 204–5.

———. "Nicomachus of Gerasa." *DSB*, 10, pp. 113–14.

———. "Parmenedes of Elea." *DSB*, 10, pp. 324–25.

Taton, René, ed. "Blaise Pascal." *DSB*, 10, pp. 330–42.

———. *History of Science.* Translated by A. J. Pomerans. 4 vols. New York, 1964–66.

Tekeli, S. "Habash al-Hasib." *DSB*, 5, pp. 612–20.

———. "Muhyil-Din Al-Maghribi." *DSB*, 9, pp. 555–57.

Thomas, Phillip Drennon. "Alfonso el Sabio." *DSB*, I, p. 122.

———. "Cassiodorus," *DSB*, 3, 109–10.

Thorndike, Lynn. *A History of Magic and Experimental Science.* 8 vols. New York, 1923–58.

Thucydides. *History of the Peloponnesian War.* Translated by Rex Warner. Harmondsworth, 1987.

Tischendorf, Constantine. *Reise in den Orient.* Leipzig, 1846. Translated by W. E. Shuckard as *Travels in the East* (London, 1847).

Tivier, Andre. "Xenophanes." *DSB*, 14, pp. 536–37.

Toomer, G. J. "Apollonius of Perga." *DSB*, 1, pp. 179–93.

———. "Campanus of Novara." *DSB*, 3, pp. 23–29.

———. "Diophantus of Alexandria." *DSB*, 4, pp. 110–19.

———. "Heraclides Ponticus." *DSB*, 15, pp. 202–5.

———. "Hipparchus." *DSB*, 15, pp. 207–24.

———. "al-Khwarizmi." *DSB*, 7, pp. 358–65.

———. "Ptolemy (Claudius Ptolemaeus)." *DSB*, 11, pp. 186–206.

———. "Theon of Alexandria." *DSB*, 13, pp. 321–25.

———. "Vitruvius." *DSB*, 15, pp. 514–21.

———, trans. and ed. *Ptolemy's Almagest.* Princeton, 1998.

Turner, Howard R. *Science in Medieval Islam: An Illustrated Introduction.* Austin, Tex., 1995.

Underwood, Paul, ed. *Kariye Djami.* 4 vols. New York, 1966–75.

Van Helden, Albert. *Measuring the Universe: Cosmic Dimensions from Aristarchus to Halley.* Chicago, 1985.

Vasiliev, A. A. *History of the Byzantine Empire.* 2 vols. Madison, Wisc., 1952.

Verbeke, G. "Simplicius." *DSB*, 12, pp. 440–43.

Vernet, Juan. " 'Abbas ibn Firnas." *DSB*, 1, p. 5.

———. "Ibn al-Baytar." *DSB*, 1, 538–39.

———. "Ibn Juljul." *DSB*, 7, 187–88.

———. "Al-Majriti." *DSB*, 9, 39–40.

———. "Al-Zarqali," *DSB*, 14, 592–95.

Vernet, Juan, and Julio Samso. "The Development of Arabic Science in Andalusia." In *Encyclopedia of the History of Arabic Science,* vol. 1, pp. 243–75.

Vitruvius: The Ten Books on Architecture (De Architectura). Translated by Morris Hicky Morgan. New York, 1966.

Vogel, Kurt. "Byzantine Science." In *Cambridge Medieval History,* new ed., vol. 4, part 2, pp. 274–79. Cambridge, 1967.

———. "Diophantus of Alexandria." *DSB*, 4, pp. 110–19.

———. "Leonardo Fibonacci (Leonardo of Pisa)." *DSB*, 4, 604–13.

Voltaire. *Letters on England.* Translated by Leonard Tancock. Harmondsworth, 1980.

Waerden, B. L. van der. *Science Awakening.* Translated by Arnold Dresden. Groningen, 1974.

Wallace, William A. "Dietrich of Freiberg." *DSB*, 4, 92–95.

———. "Saint Albertus Magnus." *DSB*, 1, pp. 99–103.

———. "Saint Thomas Aquinas." *DSB*, 1, 196–200.

Walzer, Richard. *Greek into Arabic: Essays in Islamic Philosophy.* Oxford, 1962.

Warmington, E. H. "Posidonius." *DSB*, 11, pp. 103–6.

———. "Strabo." *DSB*, 13, pp. 83–86.

Watt, W. Montgomery. *A History of Islamic Spain.* Edinburgh, 1965.

———. *The Influence of Islam on Medieval Europe.* Edinburgh, 1982.

Westfall, Richard S. *Never at Rest: A Biography of Isaac Newton.* Cambridge, 1983.

Whitfield, Peter. *Landmarks in Western Science: From Prehistory to the Atomic Age.* London, 1991.

Wiet, Gaston. *Baghdad: Metropolis of the Abbasid Caliphate.* Translated by Seymour Feiler. Norman, Okla., 1971.

Wilson, Curtis A. "William Heytesbury." *DSB*, 6, pp. 376–80.

Wilson, Leonard G. "Galen." *DSB*, 5, pp. 227–37.

Wilson, Nigel. "Archimedes: The Palimpsest and the Tradition." *Byzantinische Zeitschrift* 92 (1999): 89–101.

———. "The Archimedes Palimpsest: A Progress Report." In "A Catalogue of Greek Manuscripts at the Walters Art Museum and Essays in Honor of Gary Vikan," special issue of *Journal of the Walters Art Museum* 62 (2004): 61–68.

———. *From Byzantium to Italy: Greek Studies in the Italian Renaissance.* London, 1992.

Wipple, John F., and Allan B. Walker. *Medieval Philosophy from St. Augustine to Nicholas of Cusa.* New York, 1960.

Yates, Frances A. *Giordano Bruno and the Hermetic Tradition.* Chicago, 1991.

Youschkevitch, A. P. "Abu'l-Wafa al Buzjani." *DSB*, 1, pp. 39–43.

———. "Isaac Newton." *DSB*, 10, pp. 42–103.

Youschkevitch, A. P., and B. A. Rosenfeld. "Al-Kashi." *DSB*, 7, pp. 255–62.

———. "Al-Khayyam (Omar Khayyam)." *DSB*, 7, pp. 323–34.

Zeller, Eduard. *Outline of the History of Greek Philosophy*. 13th ed. Revised by Wilhelm Nestle. Translated by L. R. Palmer. London, 1939.

INTERNET SITES

Fermat's Last Theorem:
en.wikipedia.org/wiki/Fermat's_last_theorem

Antikythera Computer:
en.wikipedia.org/wiki/Antikythera_mechanism

Archimedes Palimpsest:
en.wikipedia.org/wiki/Archimedes_Palimpsest

INDEX

Page numbers in *italics* refer to illustrations

A NOTE ABOUT THE AUTHOR

John Freely was born in Brooklyn, New York, and grew up in Ireland and New York. He enlisted in the U.S. Navy at seventeen and was in the last two years of World War II, including duty with a commando unit in Burma and China. He went to college on the GI Bill and received his Ph.D. in physics from NYU, after which he did postdoctoral studies at Oxford in the history of science. Since 1960 he has taught physics and the history of science at Bosphorus University (formerly Robert College) in Istanbul, with other teaching stints in Athens, London, and Boston. He is the author of more than forty books, including *Strolling Through Istanbul; Strolling Through Athens; Strolling Through Venice; Istanbul: The Imperial City; Inside the Seraglio: The Private Lives of the Sultans in Istanbul; The Western Shores of Turkey; The Lost Messiah: In Search of the Mystical Rabbi, Sabbatai Sevi; Jem Sultan, The Adventures of a Captive Turkish Prince in Renaissance Europe; John Freely's Istanbul;* and *The Emergence of Modern Science, East and West.*

A NOTE ON THE TYPE

This book was set in Albertina, the best known of the typefaces designed by Chris Brand (b. 1921 in Utrecht, the Netherlands). Issued by The Monotype Corporation in 1965, Albertina was one of the first text fonts made solely for photocomposition. It was first used to catalog the work of Stanley Morison and was exhibited in Brussels at the Albertina Library in 1966.

Composed by North Market Street Graphics, Lancaster, Pennsylvania
Printed and bound by Berryville Graphics, Berryville, Virginia
Designed by Wesley Gott
Maps by Mapping Specialists, Madison, Wisconsin